SYNTHESIS OF DEEP-SEA DRILLING RESULTS IN THE INDIAN OCEAN

FURTHER TITLES IN THIS SERIES

1 J.L. MERO
THE MINERAL RESOURCES OF THE SEA

2 L.M. FOMIN
THE DYNAMIC METHOD IN OCEANOGRAPHY

3 E.J.F. WOOD
MICROBIOLOGY OF OCEANS AND ESTUARIES

4 G. NEUMANN
OCEAN CURRENTS

5 N.G. JERLOV
OPTICAL OCEANOGRAPHY

6 V. VACQUIER
GEOMAGNETISM IN MARINE GEOLOGY

7 W.J. WALLACE
THE DEVELOPMENT OF THE CHLORINITY/SALINITY CONCEPT IN OCEANOGRAPHY

8 E. LISITZIN
SEA-LEVEL CHANGES

9 R.H. PARKER
THE STUDY OF BENTHIC COMMUNITIES

10 J.C.J. NIHOUL
MODELLING OF MARINE SYSTEMS

11 O.I. MAMAYEV
TEMPERATURE—SALINITY ANALYSIS OF WORLD OCEAN WATERS

12 E.J. FERGUSON WOOD and R.E. JOHANNES
TROPICAL MARINE POLLUTION

13 E. STEEMANN NIELSEN
MARINE PHOTOSYNTHESIS

14 N.G. JERLOV
MARINE OPTICS

15 G.P. GLASBY
MARINE MANGANESE DEPOSITS

16 V.M. KAMENKOVICH
FUNDAMENTALS OF OCEAN DYNAMICS

17 R.A. GEYER
SUBMERSIBLES AND THEIR USE IN OCEANOGRAPHY AND OCEAN ENGINEERING

18 J.W. CARUTHERS
FUNDAMENTALS OF MARINE ACOUSTICS

19 J.C.J. NIHOUL
BOTTOM TURBULENCE

20 P.H. LEBLOND and L.A. MYSAK
WAVES IN THE OCEAN

Elsevier Oceanography Series, 21

SYNTHESIS OF DEEP-SEA DRILLING RESULTS IN THE INDIAN OCEAN

Edited by

CHRIS C. VON DER BORCH

Geological Research Division
Scripps Institution of Oceanography
University of California
La Jolla, California, U.S.A.

Inter-Union Commission on Geodynamics
Scientific Report No. 36

Reprinted from *Marine Geology*, Volume 26

ELSEVIER SCIENTIFIC PUBLISHING COMPANY
AMSTERDAM — OXFORD — NEW YORK 1978

ELSEVIER SCIENTIFIC PUBLISHING COMPANY
335 Jan van Galenstraat
P.O. Box 211, Amsterdam, The Netherlands

Distributors for the United States and Canada:

ELSEVIER NORTH-HOLLAND INC.
52, Vanderbilt Avenue
New York, N.Y. 10017

Library of Congress Cataloging in Publication Data

Main entry under title:

Synthesis of deep-sea drilling results in the Indian
Ocean.

(Elsevier oceanography series ; 21) (Scientific
report - Inter-Union Commission on Geodynamics ; no. 36)
"The collection of papers ... is the result of a
joint C.M.G. and I.C.G.-sponsored symposium ... held
during the 25th International Geological Congress,
Sydney, in August 1976."
"Reprinted from Marine geology, volume 26."
Bibliography: p.
1. Geology--Indian Ocean--Congresses. 2. Marine
sediments--Indian Ocean--Congresses. I. Von der Borch,
Chris C. II. International Union of Geological
Sciences. Commission for Marine Geology. III. Inter-
Union Commission on Geodynamics. IV. International
Geological Congress, 25th, Sydney, Australia, 1976.
V. Marine geology. VI. Series: Inter-Union
Commission on Geodynamics. Scientific report - Inter-
Union Commission on Geodynamics ; no. 36.
QE350.5.S95 551.4'65 78-2361
ISBN 0-444-41675-7

ISBN: 0-444-41675-7 (Vol. 21)
ISBN: 0-444-41623-4 (Series)

Printed in The Netherlands

EDITORIAL

The collection of papers contained in this special issue of *Marine Geology* is the result of a joint C.M.G. and I.C.G.*-sponsored symposium titled *Synthesis of Deep Sea Drilling Results in the Indian Ocean*, held during the 25th International Geological Congress, Sydney in August 1976. It was recognized that a considerable volume of deep sea drilling data had suddenly become available through six Deep Sea Drilling Project legs in the Indian Ocean during the early part of the 1970's, providing a huge amount of valuable stratigraphic information to add to the already voluminous geophysical and geological data in the area. Quite clearly some sort of synthesis was considered imperative and the 25th International Geological Congress seemed an ideal forum. Although the final number of participants in the resulting symposium was relatively small, due to Australia's geographic remoteness and in no small measure to available travel funds, nevertheless the papers were of highest quality. Not all participants in the symposium have published in this issue, due to other publication commitments. Three of the papers have been included despite their authors' non-attendance at the congress; they are included because of their relevance to a collection of works on the area.

The papers presented on the following pages span a wide range of relevant problems, relating particularly to the central and eastern Indian Ocean. An important work by *McGowran* is placed first, as it involves a broad biostratigraphic study of a very critical time in the framework of Indian Ocean history the early Tertiary. This philosophical paper emphasizes the synchroneity of many major oceanic and continental margin events, based on latest available biostratigraphic zonation, and presents some conclusions that should cause considerable re-thinking amongst stratigraphers and tectonicists. A paper by *Markl* follows in which an early Cretaceous magnetic anomaly sequence is identified, placing constraints on initial rifting of the western Australian continental margin and Australia's much debated relationship to India and Antarctica. *Kidd* and *Davies* follow, employing back-tracking techniques and DSDP drillhole data to establish a series of paleobathymetric and paleoenvironment maps which demonstrate the inter-relationship of sedimentation and progressive tectonic development of the Indian Ocean. The abstract by *Hsü et al.* reviews the problem of the Messinian salinity crisis in the Mediterranean and Red Sea, emphasizing the fact that the Red Sea was not connected with the Indian Ocean during the time of evaporite emplacement but joined the realm of the Indian Ocean during post-Messinian tectonic evolution. The next three papers, by *Fleet* and *McKelvey*, *Viswanatha*

*C.M.G.: Commission for Marine Geology of the International Union of Geological Sciences. I.C.G.: Inter-Union Commission on Geodynamics.

et al. and *Thompson et al.*, all deal with petrological and geochemical studies of "basement rocks" in the central and eastern Indian Ocean regions, including samples from the Ninetyeast Ridge and the surrounding ocean basins. A wealth of analytical data is included in these papers, with one unifying conclusion being that Ninetyeast Ridge basalts, which were in part emplaced subareally, differ from basalts of mid-ocean spreading ridges but resemble those of oceanic island seamounts. The final paper, by *Boltovskoy*, is placed last in sequence only because it differs from the others by being a micropaleontological synthesis, presenting what will almost certainly become the prime reference in the region for Neogene biofacies studies using benthonic foraminifera

Grateful acknowledgements are due to C.M.G. and I.C.G. for their sponsorship, for financial support for contributors, and finally for suggesting this symposium.

Chris C. von der BORCH
Geological Research Division
Scripps Institution of Oceanography
La Jolla, California
(Permanent address:
School of Earth Sciences,
Flindes University, Bedford Park,
South Australia)

CONTENTS

Marine Geology, 26 (1978) 1—39

STRATIGRAPHIC RECORD OF EARLY TERTIARY OCEANIC AND CONTINENTAL EVENTS IN THE INDIAN OCEAN REGION

BRIAN McGOWRAN

Department of Geology, University of Adelaide, Adelaide, S.A. 5001 (Australia)

(Received March 28, 1977)

ABSTRACT

McGowran, B., 1978. Stratigraphic record of Early Tertiary oceanic and continental events in the Indian Ocean region. Mar. Geol., 26: 1—39.

Primarily on the evidence of planktonic foraminifera, correlations are made among oceanic sections (northern and eastern Indian Ocean and southwest Pacific Ocean) and composite continental successions (India—Pakistan, Australian region). The paper discusses the stratigraphic patterns thus perceived and their possible meaning in terms of chronological relationship to tectonic, oceanic and climatic/biogeographic events.

In the oceans, the stratigraphy includes: hiatus across the Paleocene/Eocene boundary; a sharply defined but allochronous top to the Eocene "cherts" (early Middle Eocene, Indian; Middle/Late Eocene boundary, southwest Pacific); hiatus and sporadic accumulation of pelagic carbonates in the Middle and Late Eocene. Events in the continental record are similarly widespread: regardless of local environment one can recognize two sequences, of Paleocene—Early Eocene and Middle—Late Eocene age respectively and separated by regression, hiatus or non-identification across the Early/Middle Eocene boundary.

The stratigraphy reflects isochronous, "sudden", platewide geohistory. The available evidence indicates plate-tectonic rearrangement at about the time of the Paleocene/Eocene boundary with responses then and subsequently in the stratigraphic, climatic and foraminiferal-biogeographic record.

INTRODUCTION

Recent reviews of the Paleocene and the Eocene in the Indian Ocean region and in southern Australia concentrated on planktonic foraminiferal assemblages, biostratigraphic events and correlations, with some reference to benthonic foraminiferal biofacies and biogeography (McGowran, 1977a, b). The present paper is an outcome of those reviews. It aims to establish stratigraphic patterns in the oceans and on the continental margins, to compare the patterns, and to relate them to tectonic and climatic/oceanographic events.

The more comprehensive attempts at synthesizing Early Tertiary geohistory and biohistory tend to use a coarse Cainozoic — or later Phanerozoic — time scale. Some are cited here. One result of the present attempt at synthesis will be, hopefully, to focus attention more crisply on historical questions at the level of resolution of the planktonic foraminiferal or calcareous nannofossil zone. Thus, correlation, meaning relationship in a chronological framework, is fundamental and all data and conclusions are presented historically rather than geographically.

The "oceanic" part of this study covers the northern and eastern Indian Ocean region (Fig.1a) for which the prime source is Volumes 22—24, 26, 28 of the Initial Reports of the Deep Sea Drilling Project (DSDP), with reference to the southwest Pacific Ocean (Fig.1b) (DSDP Volumes 21, 29, 30) which, north of New Zealand, is the "Melanesian Borderlands" between the Indian and Pacific Plates (Falvey, 1975; Coleman and Packham, 1976). Transitional between "ocean" and "continent" are the Naturaliste, Queensland and South Campbell Plateaus; the Ontong Java Plateau was on the Pacific Plate, as it still is. The terms "continental" and "marginal" are used in opposition to "oceanic". General geological data are covered or cited fairly comprehensively by Powell and Conaghan (1973), Spencer (1974), Burk and Drake (1974), Anonymous (1974), Dow (1975), Leslie et al. (1976), and Douglas and Ferguson (1976). However, the essential biostratigraphic data mostly are not thus covered but are cited in McGowran (1977a, b).

CHRONOLOGICAL FRAMEWORK

The correlation charts on which this paper is based all derive from Fig.2. The calibration of planktonic foraminiferal P-Zones against geochronology is taken from Berggren (1972).

It will be apparent that the two columns of foraminiferal-biostratigraphic events have rather little in common: the best possible demonstration of biogeographic constraints on biostratigraphy. The reviews of the Paleocene—Eocene foraminiferal succession in and around the Indian Ocean (McGowran, 1977a,b) confirmed that the zonation of the Paleocene and Early Eocene is comparatively versatile, in that several of the more important members of lineages in *Subbotina*, *Planorotalites*, *Acarinina* and *Morozovella* are not markedly restricted to low latitudes. Problems in correlating assemblages in southern Australia and New Zealand are problems more of "facies" than of "latitude"; the assemblages are sifted as the environment becomes less pelagic, and the marine ingressions (ephemeral transgressions) in a marginal or paralic environment lack many species whose absence, of course, is not to be trusted in correlation.

From the base of the Middle Eocene onward, the situation seems to be reversed. Undoubtedly there are contrasts between fully pelagic and hemipelagic, or neritic, assemblages. But these contrasts are overshadowed by a sense of the increased influence of "palaeoclimate" through watermass shifts which control the primary (pelagic) distribution of species. Not one of the biostratigraphic events recognized in southern Australia above the extinction of *Morozovella caucasica* (Fig.2) plays any significant part in the definition or identification of the P-Zones (and vice versa) and yet several of these events occur in New Zealand, and probably isochronously with southern Australia.

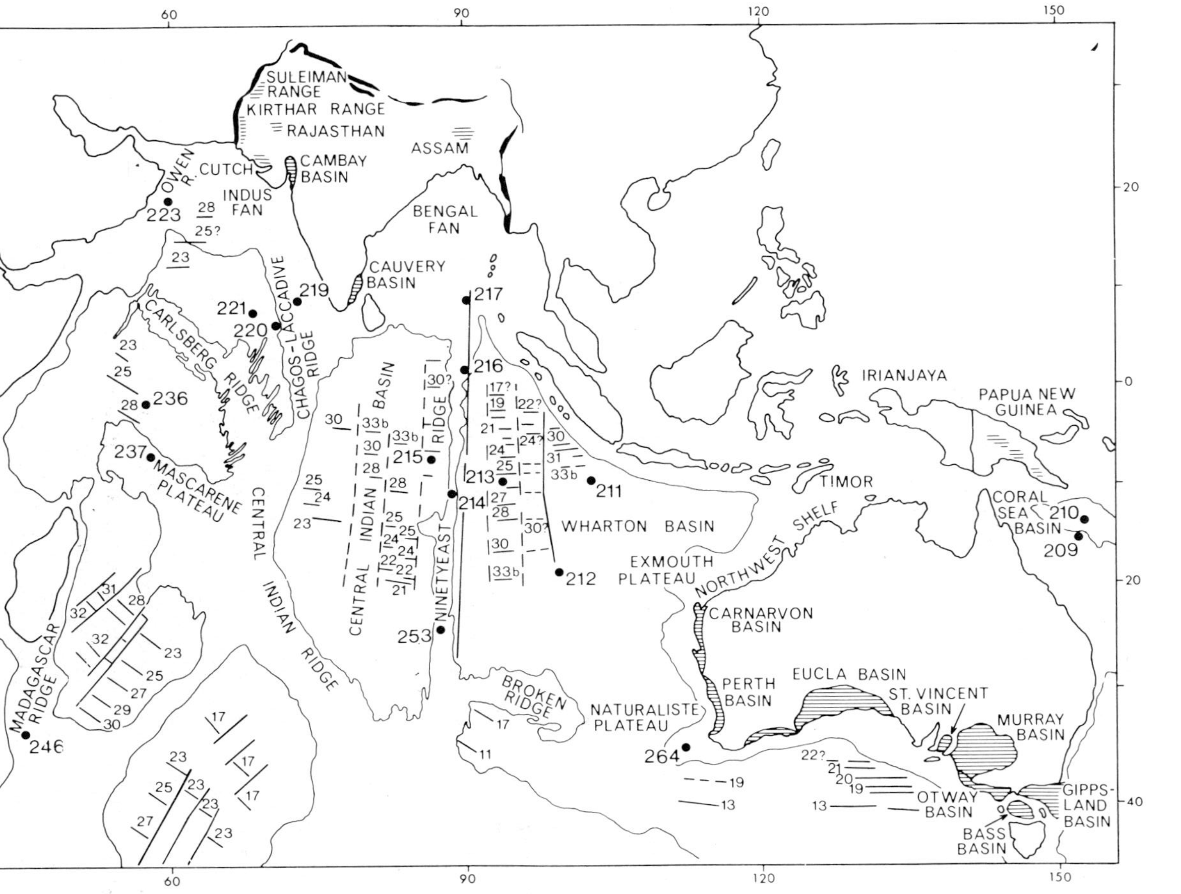

Fig.1a. Locality map, Indian Ocean Region, with DSDP Sites, geomagnetic anomalies, oceanic topography shown by 4000-m isobath. Shaded areas on continents: basins, localities or sections discussed. Indian subcontinent outlined by tectonic lineaments to show that the stratigraphy discussed here is "Indian", not "Asian". b. (p.4). Locality map, southwest Pacific Ocean, with DSDP Sites; oceanic topography outlined by 2000- and 4000-m isobaths. Based on a compilation by J.J. Veevers (see Veevers and McElhinny, 1976, fig.1).

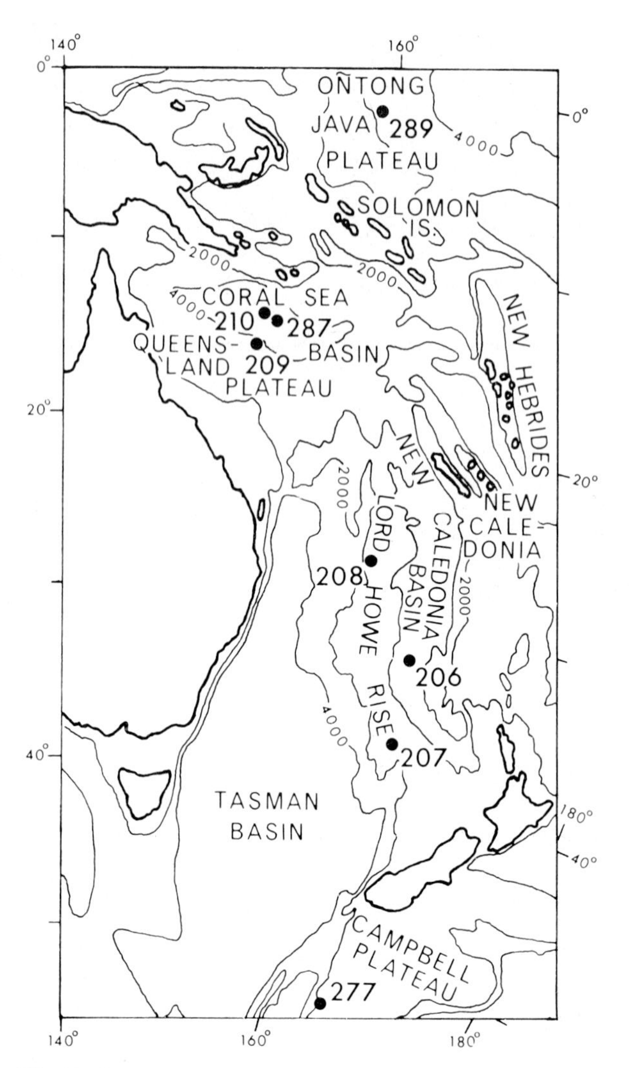

Fig.1b. For caption see p. 3.

That the contrast in biostratigraphic style between the Paleocene—Early Eocene and the Middle—Upper Eocene is a response to a reorganization of oceanic patterns foreshadows a main conclusion of this paper. However, I conclude elsewhere that this provincialism need not be expressed by the clutter of extant zonations (McGowran, 1977b). Biostratigraphic events which have been — or could be — used to define zones in extratropical southern Australia can be employed as stratigraphic markers without formal zonation. The zones are no more useful than the defining events (or "datums") and the direct correlation of "datums" with the P-Zones (Fig.2), essential but tentative, and only a best present estimate, concedes nothing in historical accuracy while contributing substantially to communication.

MA

40 -
45 -
50 -
55 -
60 -
65 -

LATE EOCENE: P.17, P.16, P.15
MIDDLE EOCENE: P.14, P.13, P.12, P.11, P.10
EARLY EOCENE: P.9, P.8, P.7, P.6b
LATE PALEOCENE: P.6a, P.5, P.4, P.3
EARLY PALEOCENE: P.2, P.1d, P.1c, P.1b, P.1a

NINETYEAST RIDGE

GLOBIGERINA TAPURIENSIS
GLOBIGERINATHEKA SPP.
TRUNCOROTALOIDES COLLACTEA
PLANOROTALITES PSEUDOSCITULA
MOROZOVELLA CRASSATA
ORBULINOIDES BECKMANNI
ORBULINOIDES BECKMANNI
SUBBOTINA FRONTOSA
GLOBIGERINATHEKA HIGGINSI
MOROZOVELLA ARAGONENSIS
TRUNCOROTALOIDES TOPILENSIS
GLOBOROTALIA CENTRALIS GROUP
GLOBIGERINATHEKA SPP.
MOROZOVELLA CAUCASICA
PLANOROTALITES PSEUDOCHAPMANI
MOROZOVELLA DENSA
MOROZOVELLA CAUCASICA
MOROZOVELLA MARGINODENTATA
MOROZOVELLA ARAGONENSIS
MOROZOVELLA FORMOSA S.S.
MOROZOVELLA LENSIFORMIS
MOROZOVELLA VELASCOENSIS S.L.
MOROZOVELLA SUBBOTINAE/ MARGINODENTATA
PLANOROTALITES PSEUDOMENARDII
PLANOROTALITES CHAPMANI
MOROZOVELLA ANGULATA GROUP
MOROZOVELLA ACUTISPIRA
PLANOROTALITES PSEUDOMENARDII
PLANOROTALITES CHAPMANI
MOROZOVELLA CONICOTRUNCATA
MOROZOVELLA ANGULATA
"SUBBOTINA" INCONSTANS
PLANOROTALITES COMPRESSA
SUBBOTINA PSEUDOBULLOIDES
SUBBOTINA TRILOCULINOIDES
GLOBOTRUNCANA, ETC.

SOUTHERN AUSTRALIA

GLOBIGERINATHEKA INDEX
TENUITELLA INSOLITA
TENUITELLA ACULEATA
GLOBIGERINA BREVIS
TENUITELLA ACULEATA
HANTKENINA PRIMITIVA
TENUITELLA GEMMA
TENUITELLA ACULEATA
TRUNCOROTALOIDES COLLACTEA
SUBBOTINA CF FRONTOSA
ACARININA PRIMITIVA
TENUITELLA ACULEATA
GLOBOROTALIA CERROAZULENSIS POMEROLI
GLOBIGERINATHEKA INDEX
PLANOROTALITES AUSTRALIFORMIS
MOROZOVELLA CAUCASICA
MOROZOVELLA AEQUA
PSEUDOHASTIGERINA PSEUDOIOTA
PLANOROTALITES CHAPMANI
PLANOROTALITES CHAPMANI GP HAUNSBERGENSIS

MAGNETIC ANOMALY

★(1): 14, 15, 16, 17, 18, 19, 20, 21, 22, 23, 24, 25, 26, 27, 28, 29

(2): 13, 14, 15, 16, 17, 18, 19, 20, 21, 22, 23, 24, 25, 26, 27, 28, 29

Fig.2. Chronological framework. Geochronological scale (Ma), chronostratigraphic scale and planktonic foraminiferal P-Zones from Berggren (1972). Composite successions of planktonic foraminiferal events on the Ninetyeast Ridge (essentially "tropical") and in southern Australia (essentially "extratropical") from McGowran (1977a, b). ⊥ first appearance, historically; ⊤ last appearance, historically; - - - marine ingressions in a marginal marine environment. Magnetic anomaly scale (Heirtzler et al., 1968) calibrated against other scales (*1*) by Sclater et al. (1974) (*used here), and (*2*) by Tarling and Mitchell (1976), modified.

Sclater et al. (1974) re-scaled the numbered geomagnetic polarity changes (Heirtzler et al., 1968) against Berggren's (1972) scale of foraminiferal biostratigraphy against geochronology (isotopic ages). With little substantiation, Tarling and Mitchell (1976) not only have altered the ages of the numbered anomalies but have changed also the biostratigraphic—geochronologic relationship. When that connexion is restored and the geomagnetic anomalies are plotted against the P-Zones as Tarling and Mitchell (op. cit.) have done (with some fudging, since the authors show biostratigraphic boundaries diachronously, to cover margins of error) then there is considerably less change between their scale and the scale of Sclater et al. (op. cit., their fig.2) than was immediately apparent. The latter scale is used here.

OCEANIC STRATIGRAPHIC RECORD

Twenty stratigraphic sections drilled during the Deep Sea Drilling Project are assembled in Fig.3. The sedimentary lithologies encountered and described in Site Reports have been reduced rather brutally to a few basic types, dominant among which are "calcareous sediment" (ooze, chalk; often with clay or, more rarely, silica) and "chert" (used here in the loose sense current in marine geology). Siliceous ooze was encountered only rarely. "Sand" is restricted to the base of the sedimentary column at four sites and includes both terrigenous and volcanogenic materials. "Clay" and silty clay at Sites 223, 210 and 287 is terrigenous; there is an important facies change from carbonate to clay in the Early Eocene at Sites 215 and 213. "Igneous rock" is included to show that the base of the sedimentary section presumably was reached; with the exception of the higher occurrence at Site 215, its relationship to the biostratigraphic and chronostratigraphic scales is, of course, not intended to be read literally.

Hiatuses are plotted — firmly or more tentatively — on the basis of the previous biostratigraphic review of the Indian Ocean sites and by using the same framework, plus the previous tropical—extratropical correlations from southern Australasia, in the southwest Pacific Ocean. Spectacular though the gaps in accumulation are, they are treated rather conservatively. The section at Site 277 is shown as continuous because the available data (Jenkins, 1975) do not, as yet, demonstrate either continuity or discontinuity in those high-latitude assemblages. Recovery at Site 289 was not good enough to resolve the pattern of accumulation and hiatus in the Zone P.11—P.16 interval, and so that interval too is shown as continuous.

The main points emerging from these correlations are summarized in order of age.

Paleocene

Earliest Paleocene was identified rarely, in part because of the hiatus at the Cretaceous/Tertiary boundary and in part because much of the oceanic crust in the region was emplaced in the latest Cretaceous to earliest Tertiary.

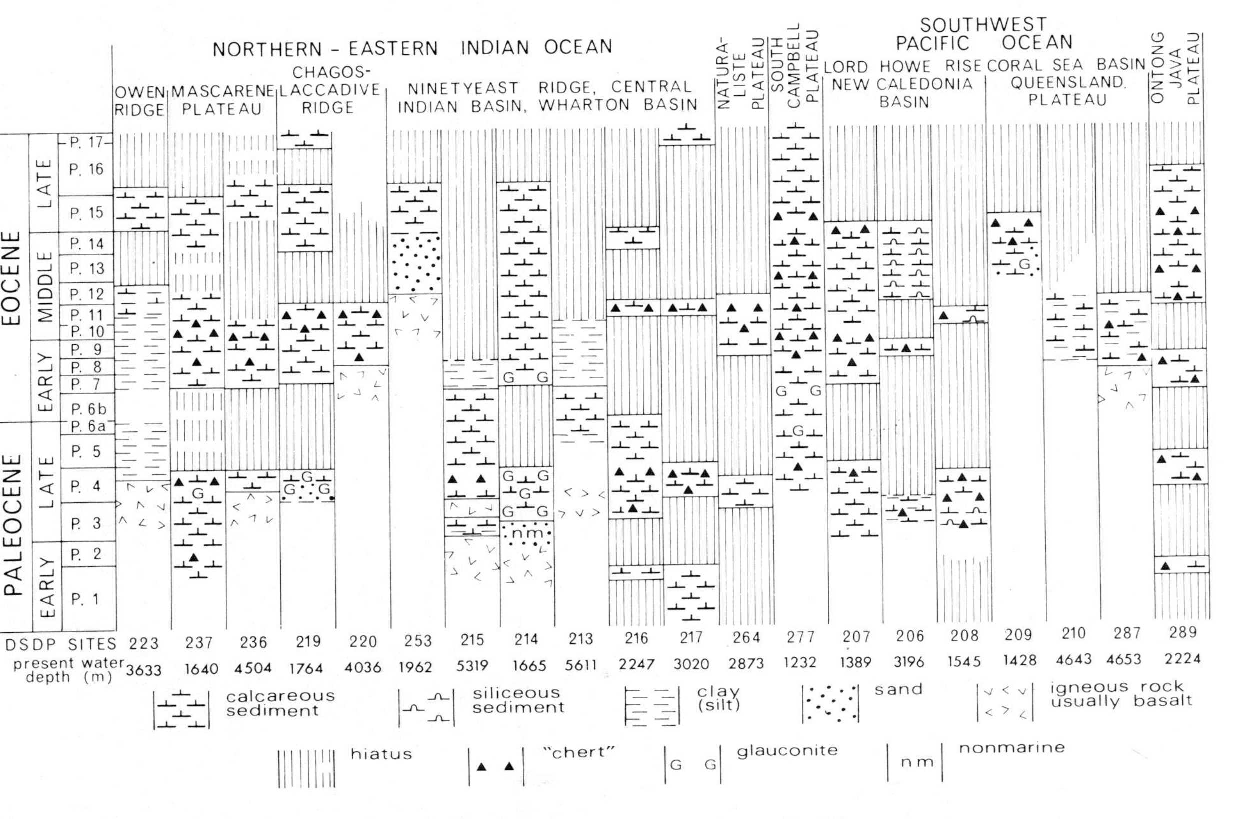

Fig.3. Stratigraphic sections at selected DSDP sites, Indian and southwest Pacific oceans.

Sedimentation was dominantly calcareous (with diagenetic cherts) in the Paleocene, becoming most widespread and consistently identified (especially on the presence of *Planorotalites pseudomenardii*) at the level of Zone P.4. Indeed, the only places where chert was not encountered were where the environment was neritic (Sites 219, 214).

Hiatus at Paleocene/Eocene boundary

The biostratigraphically determined break at this level is one of the most consistent and striking features of the oceanic record. Essentially, the evidence is the juxtaposition of Zone P.4 (with *Planorotalites pseudomenardii*) with Zone P.7 (with *Morozovella aragonensis*, at lower latitudes) or a slightly higher zone. The hiatus is virtually isochronous at six sites on topographic highs (it may be present at Site 277; Site 216 is the main exception to the generalization) but has been identified (tentatively at Site 237; McGowran, 1977a) over a wide range of present depths and palaeodepths, from neritic to bathyal (Fig.4).

In contrast, calcareous sediment accumulated during that interval in the Central Indian and Wharton Basins (Sites 215, 213) until it gave way quite rapidly to brown clay at the time of resumption of accumulation above the hiatus elsewhere, and also at the time of initial sedimentation in the Coral Sea Basin (Sites 210, 287).

Early to Middle Eocene

The pattern of calcareous sediment with chert is resumed. In the Indian Ocean, chert was encountered downhole at no less than seven of the eight sites with calcareous sections, seemingly isochronously (Zone P.11) at five, and perhaps slightly below this level at the two others where the biostratigraphy is rather less clear. In all cases the horizon is immediately below an hiatus.

The situation in the southwest Pacific is different, in that the more or less isochronous top of the cherts encountered at four sites (and siliceous ooze at a fifth, much deeper site) is in Zone P.14 or lowest Zone P.15. The top of the cherts — seismic "Horizon A" as identified pre-DSDP in the Atlantic Ocean — is not diachronous through the region as a whole but allochronous, consisting of two chronologically well-defined horizons neatly separated in space and time. McGowran (1973b) suggested that cherts in Middle Eocene carbonates in the Eucla Basin in southern Australia (shown on figs.6 and 7, op.cit.) are a continental equivalent of "Horizon A"; they are close in age to the southwest Pacific, not the Indian Ocean horizon. The lack of biostratigraphic resolution mentioned for Sites 277 and 289 precludes analysis of whether there are two successional chert—carbonate packets in the southwest Pacific (Early to early Middle, and late Middle to earliest Late Eocene) or only one, continuous in time.

Chert was not found in the calcareous sediment at Site 214 on the

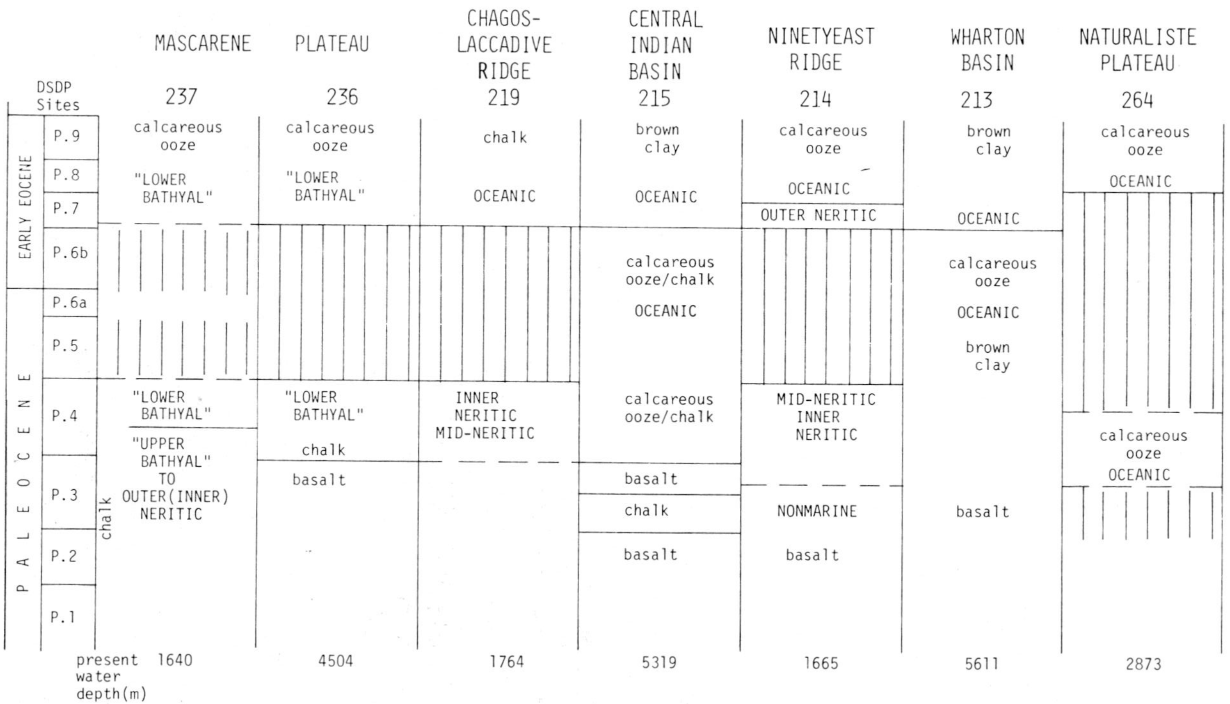

DSDP Sites

MASCARENE 237: calcareous ooze; "LOWER BATHYAL"; "LOWER BATHYAL"; "UPPER BATHYAL" TO OUTER(INNER) NERITIC; chalk

PLATEAU 236: calcareous ooze; "LOWER BATHYAL"; "LOWER BATHYAL"; chalk; basalt

CHAGOS-LACCADIVE RIDGE 219: chalk; OCEANIC; INNER NERITIC MID-NERITIC

CENTRAL INDIAN BASIN 215: brown clay; OCEANIC; calcareous ooze/chalk; OCEANIC; calcareous ooze/chalk; basalt; chalk; basalt

NINETYEAST RIDGE 214: calcareous ooze; OCEANIC; OUTER NERITIC; MID-NERITIC INNER NERITIC; NONMARINE; basalt

WHARTON BASIN 213: brown clay; OCEANIC; calcareous ooze; OCEANIC; brown clay; basalt

NATURALISTE PLATEAU 264: calcareous ooze; OCEANIC; calcareous ooze; OCEANIC

EARLY EOCENE: P.9, P.8, P.7, P.6b

PALEOCENE: P.6a, P.5, P.4, P.3, P.2, P.1

present water depth(m): 1640, 4504, 1764, 5319, 1665, 5611, 2873

Fig.4. The Paleocene/Eocene unconformity in the Indian Ocean. The evidence for an isochronous hiatus interrupting the record of rapid subsidence is summarized and discussed in McGowran (1977a).

Ninetyeast Ridge, shown in Fig.3 as a continuous section. But there is another striking correlation here. The earliest Middle Eocene (Zones P.10—earliest P.11) is not clearly resolved biostratigraphically and there could be an hiatus. There is, more clearly, planktonic foraminiferal evidence for climatic cooling (McGowran, 1974, 1977a): keeled morozovellids become sporadic in occurrence, robust acarininids and "globigerinids" are more prominent in the assemblages, and the extratropical index fossils *Planorotalites australiformis* and *Acarinina primitiva* make a brief incursion, both occupying only part of their much longer stratigraphic range as documented in southern Australasia.

Middle to Late Eocene

The record is noteworthy mainly for hiatuses, especially in the intervals of Zones P.12—P.14 and of Zone P.16. Zone P.13 (with *Orbulinoides beckmanni*) is identified only at Site 214, where planktonic foraminiferal evidence for warming (McGowran, 1977a) correlates very well with a migration southward by essentially pantropical, "larger" benthonic foraminifera (see below). Most of the identified Middle to Late Eocene carbonate is poorly preserved and the foraminiferal assemblages are dominated by robust specimens.

CONTINENTAL STRATIGRAPHIC RECORD

The eighteen columns presented in Fig.5 differ from the oceanic correlation chart (Fig.3) in two ways. The first difference is that these sections are composite, representing a summary of the succession in a sedimentary basin, or local geographic area. That is why some hiatuses are shown as diachronous, and why different facies of the same age are crammed into the same column in some cases. Such problems, however, have little bearing on the message that Fig.5 attempts to convey. Secondly, a much greater range of lithologies is found in continental sedimentary environments, partly because the environments vary more than in the deep sea but also because sampling is so much denser. The reduction of stratal variety to three symbols accordingly is more drastic and the symbols represent facies rather than lithologies.

The "mostly detrital, neritic to marginal marine" facies in southern Australia can be lignitic, usually is low in carbonate content and is characterized by marine and terrestrial palynomorphs, agglutinated benthonic foraminifera, and marine ingressions signalled by calcareous foraminifera of inner neritic aspect, low-diversity planktonic assemblages, and sporadic occurrences of shelly marine invertebrates. In Pakistan and India, regression during the Early Eocene is indicated by similar facies, but also by gypsum and shallow-water carbonates. Much of the coal of the "Laki Stage" refers to this interval, as do the pronounced changes in the diversity of foraminifera and skeletonized marine invertebrates (Sahni and Kumar, 1974).

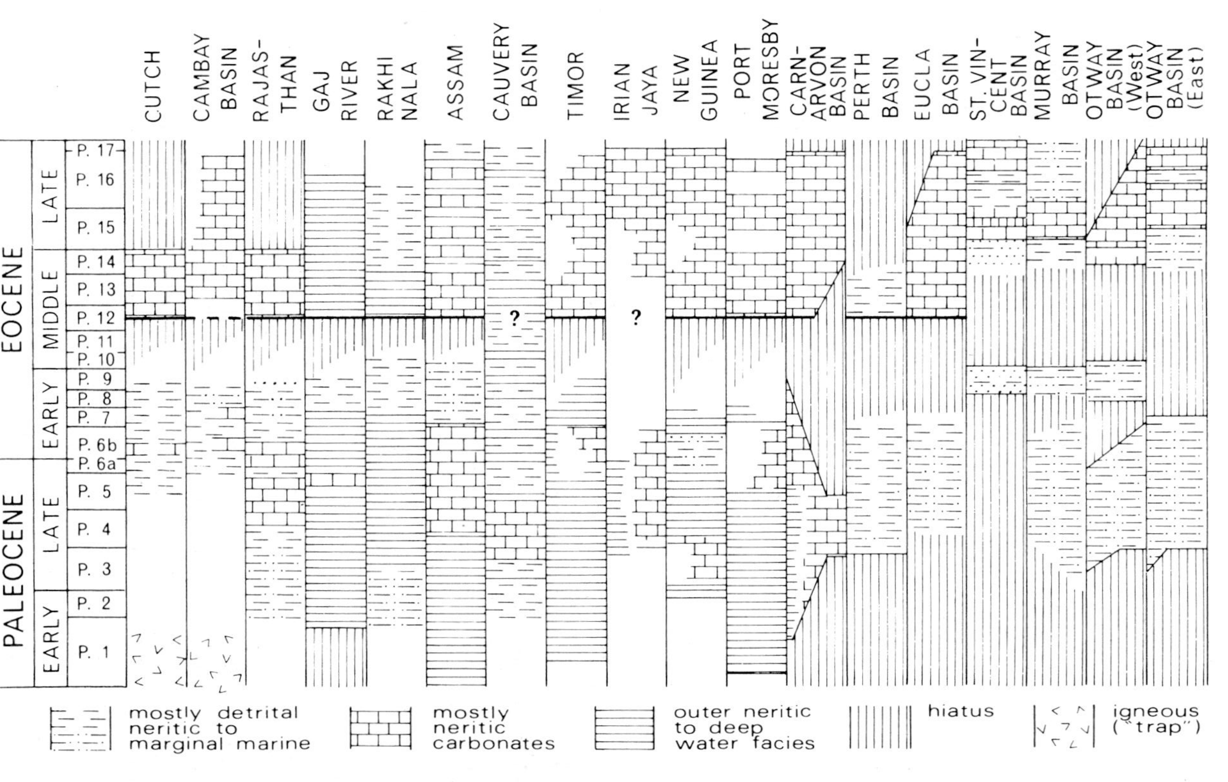

Fig.5. Highly generalized continental stratigraphic successions in the Indian and Australian regions (Fig.1). Gaj River is in the Kirthar Range; Rakhi Nala is in the Suleiman Range. See discussion in text.

The "mostly neritic carbonates" vary largely as a function of climate. In southern Australia the characteristic suite in the Middle—Late Eocene (e.g. McGowran, 1973a) includes ferruginous quartz sands and clays, glauconitic marls and greensands and bryozoan—molluscan limestones and marls. In Western Australia (in the Middle—Late Eocene, in contrast to the Paleocene—Early Eocene) and in New Guinea, Timor and Pakistan—India, this facies includes benthonic "larger" foraminifera which can be rock-forming (Adams, 1970). Platform carbonates range from "open-neritic" or "reefal", with calcareous-perforate nummulitids, discocyclinids and others, often associated with planktonics, to "lagoonal", with the porcellanous alveolinids, and others.

The "outer neritic to deep-water facies" is characterized primarily by the absence of neritic fossils (or evidence that they are displaced), often high numbers of planktonic microfossils and the presence of benthonic foraminifera of deeper-water aspect, and variation from a strong terrigenous input ("flysch") to a "starved facies" such as *Couches rouges* (reddish colour, planktonics packed together). There are now numerous localities known in a narrow zone extending from Pakistan, through Assam, Burma and the Andaman Islands to Timor, New Guinea and the Melanesian island arcs which include fine-grained limestones and lutites, sometimes cherty, often indurated and deformed and sometimes associated with volcanics and metamorphosed, with abundant planktonic foraminifera of Paleocene to Early Eocene age.

It is apparent from inspection of Fig.5, however, that the pattern of facies distribution is subordinate to a two-part distribution of *all* sediments with reference to time. There would seem to be two *stratigraphic sequences* in the sense of Sloss (1963; see also J.G. Johnson, 1971, 1974; Rona, 1973) except that Sloss distinguished six sequences spanning the Phanerozoic on the North American craton whereas there are two here, within the Palaeogene alone. These packets of sediment, which are neither lithostratigraphic, biostratigraphic nor chronostratigraphic and which are not recognized in stratigraphic codes, are labelled as Sequence One and Sequence Two in Figs.6 and 7. There are four sequences in the Australian Cainozoic (McGowran, in preparation).

Sequence One: Paleocene and Early Eocene

Sedimentary accumulation began in the Zone P.1—P.2 interval in the deeper-water sections of the active margins and offshore Western Australia; earliest Danian Stage fossils are reported only from Assam and the Andaman Islands. On passive margins and over the cratons the Sequence begins rather later. But in all tectonic and sedimentary environments there appears to be a maximum in Zone P.4: *Planorotalites pseudomenardii* which defines Zone P.4 is the most ubiquitous index fossil of all Palaeogene planktonic foraminifera, and all the available evidence indicates that pre-Zone P.4 occurrences of neritic "larger foraminifera" are insignificant (Fig.8). Zones P.6 and P.7 or their equivalents — including the marine ingressions in southern Australia (some are labelled with fossil names in Fig.2) — have been identified in numerous reports, but the planktonic foraminiferal sequence comes to a halt in the later part of the Early Eocene. In some cases (India and Pakistan, Western Australia, southern Australia) disconformity can be identified in actual stratigraphic sections from the absence of most or all of Zones P.8—P.11 and from facies changes and reworking indicating regression. In other cases, especially the New Guinea region, the same conclusion is reached from a more general accumulation of evidence: planktonic foraminiferal dating of assemblages of free specimens and of thin-sections of indurated sediments give a range from Zone P.1 to P.6 or P.7, but Zones P.8—P.11 seem to be virtually absent. The stratigraphic control on the geology of Irian Jaya (Visser and Hermes, 1962, Enclosure 7 and text) is revealing: in the Eocene there is only one "time stratigraphic unit" between the *Globorotalia velascoensis* Subzone (= Zone P.6a) and the incoming of Zone Tb larger foraminifera (= Zones P.15—17, possibly Zone P.14). Finally, it is noteworthy that Late Cainozoic orogenic sediments on or derived from the active plate margin contain reworked Paleocene to Lower Eocene planktonic foraminifera rather commonly, whereas Middle Eocene and younger specimens are much rarer, or absent. Examples from the writer's experience of this generalization include the Bengal Fan, Timor, and the New Guinea Highlands. Reworking began early in the Coral Sea region where Early Paleocene species were found in the Early Eocene at Site 287 (Site Report; Andrews et al., 1975).

Sequence Two: Middle Eocene and Late Eocene

The hiatus, regression, or stratigraphic non-identification extends into the earlier part of the Middle Eocene. The Late Eocene carbonates of the Indo-Pacific region, recognized as a major stratigraphic event for half a century and more, begin with a transgression in the Middle Eocene. This is the Khirthar transgression in the Pakistan—India—Burma region (Nagappa, 1959). It seems now to have been isochronous, within the limits of biostratigraphic resolution, around the entire margin of the northern and eastern Indian Ocean. (Fig.5 is conservative: perhaps the bold line should be extended to the Otway Basin, but correlation of the extratropical assemblages is not yet adequate.) Site 209 on the Queensland Plateau (Fig.3) encountered but did not drill through a neritic-type section in the Middle Eocene which may well have been part of the continental Sequence Two before pronounced subsidence admitted an oceanic environment.

A second contrast with Sequence One is the greater prominence of platform carbonates. Bryozoan limestones and marls now accumulated in southern Australia and their tropical equivalents are more widespread than they were previously. In Papua New Guinea, the Maastrichtian, Paleocene and Early Eocene are confined to a narrow belt whereas the Late (and now, undoubtedly, later Middle) Eocene was extensive over the platform to the south (though now patchy, after Oligocene erosion). The biostratigraphic resolution of the Sequence Two carbonates is poor except where free planktonic foraminiferal specimens are obtained; "larger" foraminifera divide the interval of Zones P.12—P.17 only into Zones Ta_3 and Tb in the Indo-Pacific letter classification (Adams, 1970). It is possible, but not yet resolvable, that there is a significant event in the vicinity of the Middle/Late Eocene boundary; there is some evidence for recognizing a Middle Eocene and then a more far-reaching Late Eocene transgression in southern Australia.

In broad terms Sequence Two is terminated by hiatus close to the Eocene/ Oligocene boundary. In biostratigraphic terms, Zones P.15—P.17 in toto are identified more commonly than is Zone P.18 and Zone Tb likewise is more common, and more diverse in its assemblages, than is the Early Oligocene Zone Tc (Adams, 1970).

OCEANS AND CONTINENTAL MARGINS: COMPARISON AND CONTRAST

The data of Figs.3 and 5 are condensed further for direct comparison in Fig.6, which thus is an impressionist rendering of the stratigraphic style. There are three quite clear generalizations which do not depend on the highly qualitative judgments involved in deriving Fig.6:

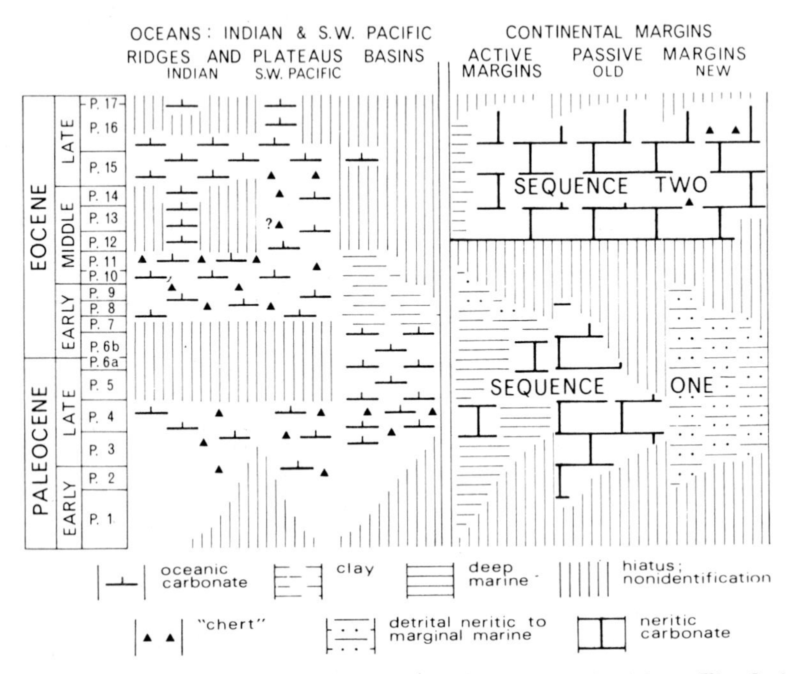

Fig.6. Summary of the stratigraphic generalizations derived from Figs.3, 4, and 5. "Active" continental margins refers to, especially, Timor and the island of New Guinea; "passive", primarily to Western Australia ("old") and southern Australia ("new"); the passive margins of India, although "old" in comparison with southern Australia (Anomaly 22) are more similar to that margin (sedimentary facies; volcanics in Fig.1) than to the Western Australian margin.

(1) The pattern of accumulation in the Paleocene is rather similar in the oceans and on and marginal to the continents in beginning in Zones P.1—P.2 and in maximizing in Zone P.4. On the margins, furthermore, variations in facies and tectonic environment seem to be overridden in the development of Sequence One.

(2) The two main times of hiatus in the oceans — Paleocene/Eocene boundary and later Middle Eocene — are out of phase with the continental regression and hiatus across the Lower/Middle Eocene boundary. Sequence One was concluding as calcareous accumulation resumed in the Early Eocene on the oceanic structures and as carbonate gave way to clay in the Central Indian and Wharton Basins.

(3) The Middle Eocene transgression (Zones P.12—P.13) and the accumulation of carbonates as Sequence Two on the continents occurred as the carbonate record in the oceans became poor, with numerous records of hiatus above Zone P.11.

STRATIGRAPHY, TECTONICS AND CLIMATE: TOWARDS A GEOHISTORICAL SYNTHESIS

The correlations discussed thus far are extended in Fig.7 to include a variety of data from the geological, geophysical and palaeontological record. The following discussion intends to demonstrate that the stratigraphy records a rather close relationship in time between geohistorical events as diverse as plate-tectonic rearrangement, foraminiferal biogeography and the distribution of oceanic cherts.

Chronology and timing of tectonism

Sea-floor spreading. The evolution of the Indian Ocean in the region between India, Australia and Antarctica has taken place in three fairly distinct phases. From the Middle—Late Jurassic to the Late Cretaceous, the spreading ridge had a northeast—southwest trend; perhaps there were major events forming oceanic crust off northwest Australia in the Jurassic and between Greater India and Antarctica in the Early Cretaceous (the work of Veevers and his colleagues on this topic is cited and summarized in Veevers and McElhinny, 1976). At approximately 75—80 Ma (pre-Anomaly 33) a spreading system striking east—west took over with the development of enormous north—south fracture zones (Sclater and Fisher, 1974). This system lasted through the Eocene until the present system with its northwest-southeast strike at about the time of Anomaly 13 became dominant (Sclater et al., 1976). The history of the middle part of this three-part succession — the episode of south—north spreading — is of concern here. In particular, Anomaly 22 seems to mark a natural subdivision of the Anomaly 33—13 interval.

During the Maastrichtian and Paleocene spreading was very rapid, with India moving rapidly northward (Sclater and Fisher, 1974). The Ninetyeast Ridge was on the Indian Plate and there is very good stratigraphic evidence

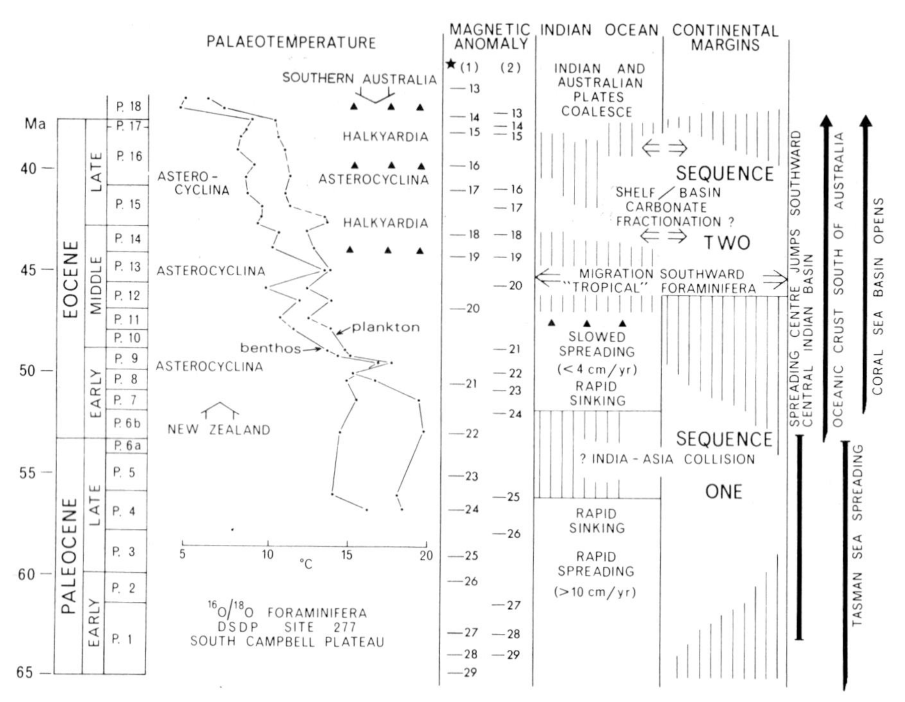

Fig.7. A historical summary of oceanic, continental, climatic and tectonic events in the Early Tertiary, derived from previous figures and discussion; for sources, see text. "Chert" symbols as in Figs.3 and 6.

for the concomitant rapid sinking to be expected: in the Late Campanian to Maastrichtian (Site 217), Maastrichtian (216) and Paleocene (214). The Ridge sank as it moved northward and the homotaxial succession of facies is diachronously younger to the south (Pimm et al., 1974; McGowran, 1977a).

At the end of the Paleocene (Anomaly 22, 53 Ma) several things happened at about the same time. There was a major adjustment to the system in the eastern Indian Ocean, where the spreading centre jumped southward some 11° between Anomalies 28 and 22 (Sclater et al., 1976). Australia separated from Antarctica (Weissel and Hayes, 1972). Spreading in the Tasman Sea, which began at the time of the 75—80 Ma readjustment, ceased (Hayes and Ringis, 1973). On stratigraphic evidence from Sites 210 and 287 (Fig.3) the Coral Sea began to open (Andrews et al., 1975).

During the Eocene, after the clustering of events at Anomaly 22, spreading rates in the eastern Indian Ocean were markedly less than before (Sclater and Fisher, 1974). At that time India, Antarctica and Australia were on separate plates; the Indian and Australian plates coalesced during the change to the present system. Whitmarsh et al. (1974) detected palaeomagnetically a two-phase history in the northward movement of the Indian Plate ". . . at a mean rate of 26 cm/yr from the end of the Cretaceous until at least Middle Eocene times, followed by a slower mean rate of drift of 16 cm/yr beginning in the

Miocene." But it is doubtful whether their data demand that termination of the first phase be as late as Middle Eocene.

Active plate margins. There are several items of evidence suggesting that the critical time of Anomaly 22 and the Paleocene/Eocene boundary in the oceans is also a time of some sort of climax at the active margins. The evidence comes from both the Indian and the Australian margins.

Speculation (see Powell and Conaghan, 1975) about the evolution of the Himalayan orogenic belt can be classified: (1) whether it was a continent/continent collision (which covers most plate-tectonic models), or an intra-continental phenomenon (Crawford, 1974), or whether it requires no continental drift at all (Karunakaran et al., 1976; all recent literature ignored); and (2) in terms of the sequence of discerned events, or stages, including early orogenic history, a mid-Miocene climax and Late Cenozoic uplift. Recent discussion of an Early Tertiary event, perhaps an initial collision, is of interest here. To Powell and Conaghan (1975) the magnetic anomalies in the Indian Ocean (Sclater and Fisher, 1974) indicate a collision at the time of Anomaly 23, when northward motion of India was almost halted; their geological constraint (Powell and Conaghan, 1973) is that the Middle Eocene (Khirthar) transgression might have occurred in response to post-collision isostatic readjustment. According to Stonely (1974), in Pakistan "there is no evidence to suggest contact between the Eurasian and Gondwana continents before the early Eocene: at that time, however, granitic detritus apparently derived from north of the suture flooded onto the northern portion of the ('Gondwanan'; southern) miogeosyncline." D.G. Moore et al. (1974) and Curray and Moore (1974) stress the significance of an extensive regional unconformity under the Bay of Bengal, recognized as a seismic reflection surface and as a velocity contrast, and dated by extrapolation to the Paleocene/Eocene unconformity at Site 217 on the Ninetyeast Ridge (see fig.2, op.cit.). For these authors, the regional unconformity reinforces the oceanic geomagnetic evidence (above) for a change in relative plate motion. They date the initial India/Asia contact as latest Paleocene and interpret the unconformity as some, not clearly understood, mid-plate response to an event at the active plate margin, and marking a natural division between the younger "modern" Bengal Fan and the older Cretaceous to Paleocene packet of terrigenous, detrital sediment which is part of the record preserved around a large part of the active margin (their Indus Flysch, Naga Hills Flysch, Indoburman Flysch, Andaman Flysch).

The distribution of terrestrial placental mammals might be significant. Colbert (1973) notes that the Eocene mammals known from the region are outside peninsular India; they occur in Burma and biogeographically are Asian, not Indian, and India is a blank in the Paleocene and Eocene. But recent finds of Middle Eocene age in Cutch and in Pakistan convince Sahni and Kumar (1974) that there was sufficient contact with terrestrial Asia to permit migration, perhaps by island-hopping, by earliest Middle Eocene time at the latest.

In Timor, Audley-Charles and Carter (1972) date a Timorean orogenic phase between Late Maastrichtian and earliest Eocene limits. The implication that the Australian margin was directly involved was withdrawn subsequently, the evidence now being interpreted as having been carried onto Timor during Pliocene thrusting (Carter et al., 1976), but these authors cite references to widespread, late Cretaceous to mid-Eocene tectonism in the Outer Banda Arc of Indonesia.

An ophiolitic suite of rocks outcropping through the island of New Guinea includes, most prominently, the Papuan Ultramafic Belt. In a "failed or choked subduction model" (H.L. Davies and Smith, 1971; H.L. Davies, 1976) the failure of a north-dipping subduction zone is dated as earliest Eocene on a K—Ar date of 52 ± 1.2 Ma from the thrust zone where oceanic and continental crust are now in contact. An alternative viewpoint is that this event occurred in the Oligocene (Dow, 1975) partly on the evidence of dates from both parts of the paired metamorphic belt now recognized (Ryburn, 1976). A preferred age of 50 Ma was obtained for a metamorphic event in the Solomon Islands (Richards et al., 1966). McKenzie (1975) has found that "a major shift in volcanic activity . . . took place in the Paleocene—early Eocene. Volcanism ceased in 'mainland' Papua New Guinea and commenced in the ancestral Torricelli—Finisterre—New Britain chain some distance to the northeast . . .".

Significance of the oceanic Paleocene/Eocene unconformity. The preceding discussion shows that there is a correlation in time between plate-tectonic changes inferred from oceanic geomagnetic data and individually tenuous but cumulatively more impressive items of tectonism at the active margins of both the Indian and the Australian Plates. The unconformity on oceanic ridges and plateaus defined in Figs.3 and 4 occurs at just that time. Such a good correlation invites a single tectonic or geodynamic explanation of the entire set of observations.

Pimm and Sclater (1974) and T.A. Davies et al. (1975) have drawn attention to the Paleocene/Eocene hiatus. Pimm and Sclater (op. cit.), considering only the Ninetyeast Ridge sites, suggested that the hiatus was "related solely to local tectonism". T.A. Davies et al. (op.cit) ascribed it to oceanic changes deriving from climatic deterioration and Antarctic glaciation — an explanation which now seems unlikely (see below). For Site 219, Weser (1974) proposed a local explanation of sedimentary bypass and/or erosion.

In the Late Paleocene the Mascarene Plateau was on a different plate from the Chagos—Laccadive Ridge and the Ninetyeast Ridge, and the Naturaliste Plateau was on a third plate. Other important aspects of the hiatus are its range in depth (Fig.4), its distribution from neritic-type, glauconitic sediments, to pelagic carbonates, to the terrigenous turbidites of the Bengal Fan, and its restriction to the oceans: although the record above Zone P.4 is not entirely clear, the hiatus is not clearly visible on the continental margins, active or passive. In this respect it contrasts strongly with the Cretaceous/

Tertiary and Early Oligocene hiatuses. Extensive mass wasting of oceanic sediment seems unlikely; there is only some evidence of up-section, down-slope reworking of foraminifera at Sites 237 and 219 (cited in McGowran, 1977a). It seems best to ascribe the hiatus to tectonism, more pervasive and more fundamental than acknowledged so far (D.G. Moore et al., 1974; Curray and Moore, 1974; Pimm and Sclater, 1974) but not understood beyond recording a plate-tectonic change of considerable magnitude. (It is noteworthy that D.G. Moore et al. did not obtain a reliable estimate of the hiatus recorded in the unconformity in the Bengal Fan by extrapolating to Site 217 because that section has a sampling gap where later Early Eocene sediment might be expected. Also, the hiatus is now defined more precisely in biostratigraphic terms — Zones P.5—P.6 — than it is in the charts of T.A. Davies et al. (1975).)

Early Eocene and the gap between the marginal sequences. Discussion in the literature of the timing of tectonic events tends to resolve into a choice between two or more rather distinct intervals of time. The emplacement of the Papuan Ultramafic Belt and associated metamorphism seems to be either near the Paleocene/Eocene boundary or in the Oligocene, when it might correlate with obduction in New Caledonia (Coleman and Packham, 1976). An alternative for the India/Asia collision is (although not substantiated) between about 43 and 39 Ma (Sclater et al., 1976). The evidence and the speculation in the literature suggest that the intervening time interval is rather bare of oceanic and marginal tectonic events.

The resumption of sedimentary accumulation in the Early Eocene on plateaus and ridges in the Indian Ocean is accompanied by rapid subsidence of the ocean floor, and this correlates excellently with the change from carbonate to clay at the two deepest basin sites (Fig.4). That correlation points to deepening of the ocean rather than to shoaling of the CCD. The most spectacular example of subsidence is at Site 219; in the Paleocene the site was neritic and at about the same level as peninsular India nearby, whereas it had reached bathyal depths by the Early Eocene so that India and the Chagos—Laccadive Ridge were structurally independent (Weser, 1974). If, as discussed above, decrease in spreading rates in the Indian Ocean is a response to India/Asia collision, then a general oceanic subsidence due to crustal cooling would be expected, and that expectation is supported strongly by the oceanic and continental stratigraphic record.

A casual connection between the time-correlated termination of marginal Sequence One and the relatively passive subsidence of the Indian Ocean basins and highs thus is very likely. Rapid plate motion — i.e. lithospheric accretion — and increased size of the oceanic rift bulge cause displacement of water and relative continental submergence, or a "thalassocratic" state (e.g. Hays and Pitman, 1973; Fischer, 1974). This was the situation in the Paleocene (Zone P.4). The converse, causing emergence or an "epeirocratic" state, was the situation in the later Early Eocene. Clearly, one cannot restrict this continent/ocean matching to one region since both plate-tectonic and tectono-eustatic changes are global (e.g. Rona, 1973, 1974). The precision in correlation attempted here has still to be attempted at the global scale. However, it might be that a change in spreading rates in the vicinity of 50 Ma (e.g. Vine, 1973; Flemming and Roberts, 1973; Rona, 1973; all using coarse time scales) is pointing to something fundamental within the Early Eocene.

The termination of Sequence One includes not only regression and hiatus on passive margins but the temporary disappearance from the identified record of outer neritic to pelagic facies on the active margins. That this tectono-eustatic event is perceived in the island of New Guinea and in Timor implies some form of tectonic consolidation (and that inference, incidentally, supports more "direct" reasons for a preceding orogenic episode). The Eocene was the time of conclusion in Papua New Guinea of the "Mesozoic to Tertiary Geosyncline" which was succeeded stratotectonically by the "Cainozoic Papuan Geosyncline" whose record begins in the Late Oligocene (Brown et al., 1975). Discussions of tectonic evolution in New Guinea lack chronological precision in the Palaeogene, but Brown et al. (1975, especially fig.8) show two great unconformities, pre- and post-Late (now including later Middle) Eocene, with the lower unconformity crossing all tectonic environments. It is now possible to propose that an episode of tectonism and consolidation occurred in the Early Eocene; certainly the evidence would be bolstered by a confirmation of that data for emplacement of the Papuan Ultramafic Belt, but the data are too diverse to actually need such a confirmation.

As if to anticipate and refute the theme of this discussion, Hermes (1974) stresses that in Irian Jaya there were "no discrete phases in the sense of short periods of mobility alternating with long periods of quiescence." At the scale of resolution attempted here, the stratigraphic record in Irian Jaya does not disprove an Early Eocene event. Hermes also states that "in the deeper water of the eugeosyncline deposition of extrusives with intercalations of radiolarites and pelagic limestones went on in Paleocene and Eocene times. In the Late Eocene, however, there also occur shallow-water limestones with larger foraminifera in a few localities, which are taken to show that probably tectonic activity had started in the eugeosyncline by this time." It is more likely that there was a stratotectonic succession, with neritic Sequence Two carbonates accumulating on mostly deep-water Sequence One rocks which meanwhile had accreted to the Australian continent.

There is, then, no simple contrast in terms of transgression and regression, or of sediment and hiatus, when one compares mobile belts and associated platforms on active margins with passive margins. Isochroneity and the perception of sequence are intercontinental and extend throughout the available record regardless of tectonic or sedimentary environment.

Unknown tectonic significance of Sequence Two. Between the Early Eocene and the Early Oligocene the most prominent geological phenomenon clearly is the widespread Middle Eocene transgression. This event has had several explanations, all of them "local" and therefore inadequate (see below). It is, certainly, a tectono-eustatic and stratotectonic marker, and yet plate-tectonic and other tectonic events in the vicinity of Anomaly 20 are not at all prominent in the literature. One change in the region is migration to the south of the Pacific—India pole of rotation (Coleman and Packham, 1976, fig.8).

The Early Oligocene regression terminating Sequence Two occurs in the oceans and on the continents and has been subjected to considerable palaeoclimatic and palaeo-oceanographic explanation (e.g. T.A. Davies et al., 1975; Kennett et al., 1975; Kennett and Shackleton, 1976; references cited therein). But this was a time of plate-tectonic change both east and west of Australia. Coleman and Packham (1976) correlate formation of the Caroline and South Fiji Basins with obduction on New Caledonia, possible temporary cessation of subduction, and temporary relocation of the Pacific—"India" plate boundary. In the Indian Ocean, the old spreading system gave way to the present system and the Indian and Australian plates coalesced (Sclater et al., 1976). The regression and the profound changes in the oceans and in climate (see below) may well have been triggered tectono-eustatically.

Climatic and oceanographic changes

We are now in a position to discuss climatic trends in the Early Tertiary. The topic challenges tectonics in its ability to highlight the best and the worst in geohistory (and biohistory), from the critical focus gained in cross-referencing large amounts of diverse evidence to a bland disregard for the quality of that evidence — pre-eminently, as always, the accuracy and precision of correlation. As in the previous discussion, this section stands or falls on the reliability of the chronology.

Palaeoclimatic trends. Shackleton and Kennett (1975) have produced from Site 277 the only oxygen isotope curves available from closely spaced samples under reasonable biostratigraphic control in the interval from Late Paleocene to Early Oligocene. There are sampling gaps at critical times in the curves published by Savin et al. (1975) whose data are much stronger in the Oligocene and Neogene. The South Campbell Plateau was at a high latitude in the Palaeogene and the foraminiferal assemblages show this clearly (Jenkins, 1975). Shackleton and Kennett did not scale their Site 277 curve against time. I have attempted that exercise elsewhere (McGowran, 1977b) as an outcome of identifying homotaxial planktonic foraminiferal biostratigraphic events between southern Australia, New Zealand and Site 277, and correlating the horizons (considered to be isochronous) with the P-Zones (Fig.2). That curve is reproduced here as part of Fig.7. Biostratigraphic control is weak in the Late Paleocene to earliest Eocene but that is not disastrous; higher, the control seems to be good enough for the main results to be reliable. The isotopic results can be summarized as: (1) overall temperature decrease; (2) a rather dramatic decline from the Early to the Middle Eocene and second decline across the Eocene/Oligocene boundary (or, just within the earliest Oligocene, as Kennett and Shackleton (1976) now believe, on slender evidence); (3) reversals showing as warm spikes, especially in Zones P.9 and P.13 and in the earliest Late Eocene. Caution on the interpretation of oxygen isotope ratios (e.g. Berger and Roth, 1975) is acknowledged, but the exercise here is the fit with other evidence, with chronology as the point of departure.

Biogeographic evidence: "larger" foraminifera. Tertiary foraminiferal distribution in space and time in southern Australia reflects an extratropical environment which varies according to climate, so that one can perceive the tropical planktonic (Berggren, 1972) and "larger" benthonic (Adams, 1970) biostratigraphic record through "windows", as it were, when the climatic belts expanded briefly to the south (McGowran and Wade, 1967). However, a review of the actual occurrence of the "larger", pantropical genera of the Indo-Pacific Letter Zones Ta_1, Ta_2, Ta_3 and Tb and of the analogous Maastrichtian assemblages in terms of planktonic foraminiferal zones showed that, even in the tropical platform carbonate facies, the assemblages occur in three rather distinct intervals, respectively of Maastrichtian, Late Paleocene to Early Eocene, and Middle to Late Eocene age (Fig.8). This distribution is partly a matter of evolutionary radiation and extinction and partly a reflection of Sequences One and Two (and their Maastrichtian equivalent), the latter effect stimulating the former by controlling the distribution of suitable neritic environments.

The other main points emerging from Fig.8 are:

(1) In the Maastrichtian and in Sequence One the "larger" foraminifera are restricted to the tropics except in the western Indian Ocean, thus reflecting a counterclockwise gyre (McGowran, 1968). There were neritic carbonate environments in the Carnarvon Basin and on the Ninetyeast Ridge, but there is no evidence that they were ever occupied.

(2) The time of the Khirthar transgression in the Indian region and indeed the general onset of Sequence Two neritic carbonate accumulation is the time also of rapid southward migration — on the Ninetyeast Ridge to Site 253 as well as in Western Australia. The Late Eocene migration was more extensive, reaching central-southern Australia. As a general rule, New Zealand was more accessible than was southern Australia throughout the Tertiary.

These migrations are in fact short-lived excursions. Three of the five excursions to southern Australia and New Zealand distinguished here (Fig.7) correlate well with the temperature curve. The event in Zone P.9 matches also the widespread distribution of *Morozovella caucasica* in the southern extratropical region and the last marine ingression in Sequence One in southern Australia. In Zone P.13, there is planktonic foraminiferal evidence at Site 214 for an episode of warming (McGowran, 1977a). Above the event at the beginning of the Late Eocene the curve does not resolve a series of quite sudden and reversible lithostratigraphic and biostratigraphic events which probably are synchronous across southern Australia (McGowran, 1977b). The *Hantkenina* interval (shown only as a top-occurrence in Fig.2) is close to the *Asterocyclina* occurrence at the Zone P.15/P.16 boundary.

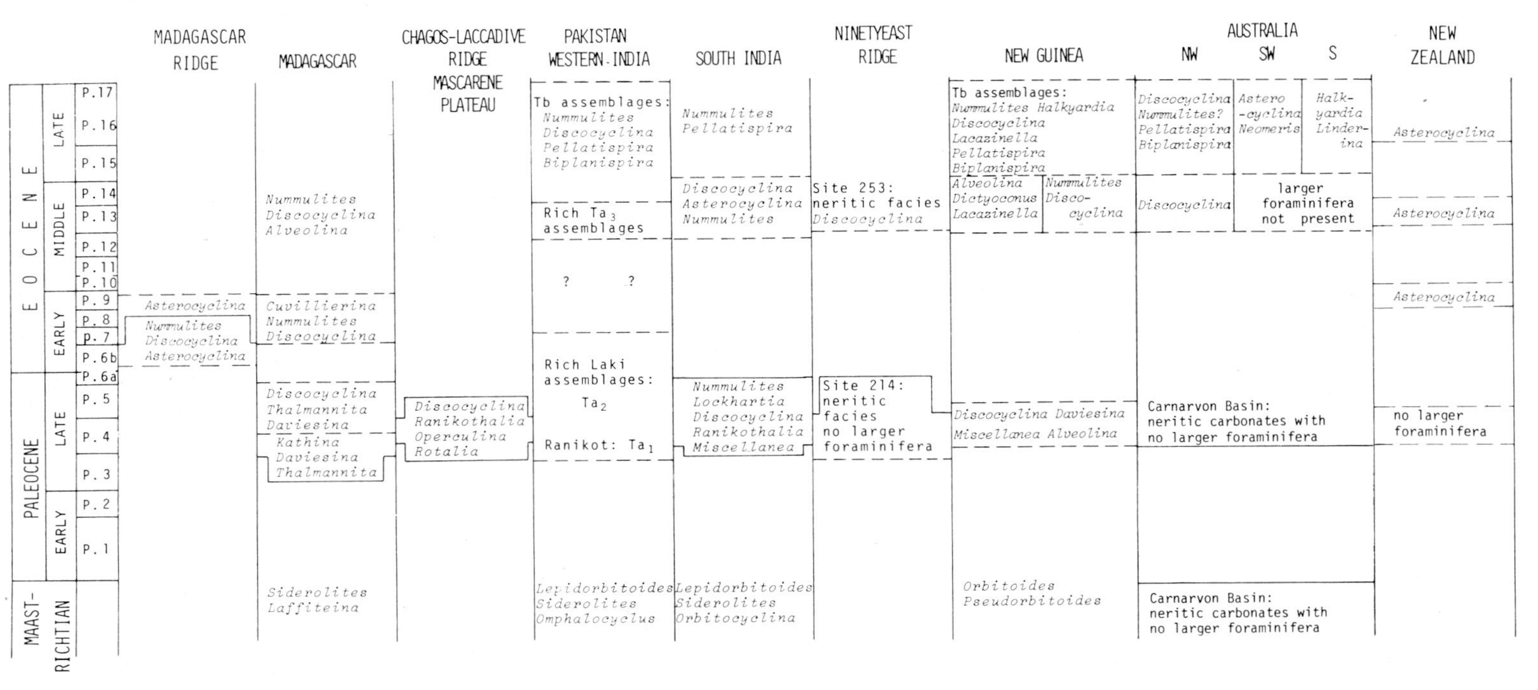

Fig.8. Distribution of pantropical, "larger" benthonic foraminifera. Correlation of assemblages is discussed in McGowran (1977a, b). The rich tropical assemblages of the Late Paleocene—Early Eocene and Middle—Late Eocene are not shown comprehensively (for an excellent summary, see Adams (1970)). Note (1) the three-part distribution of assemblages, in terms of time; (2) the more extensive occurrences in the Middle—Late Eocene compared to the Maastrichtian and the Late Paleocene.

In summary, tropical-type "larger" benthonic foraminifera were restricted in their spatial distribution during the Maastrichtian—earliest Eocene, before the climatic deterioration shown in the isotope curve. After that decline, there were far-ranging excursions out of the tropics in the Middle and Late Eocene even though the isotope curve shows no restoration to the earlier temperature values. The excursions were not "local"; they occurred in mid-ocean (Site 253) and against a counterclockwise gyre in Western Australia, and the records extend from the western Indian Ocean to New Zealand. They represent a fundamental change in the oceanic regime and their evidence may be counterpointed by the distribution of marine non-detrital silica (below).

Terrestrial floras. In the early Tertiary record of terrestrial palynomorphs in southern Australia (Otway Basin), Harris (1971) recognizes a pronounced microfloral change within the Early Eocene (between his *Cupanieidites orthoteichus* and *Proteiacidites confragosus* Zones). This change is coeval with Zones P.7—P.8 and therefore correlates with the beginnings of the isotopic temperature decline. The *P. confragosus* Zone includes a remarkable, wet forest flora (Lange, 1970) of some biogeographic significance at a high palaeolatitude (Martin, 1975); the flora coincides in age with the isotope temperature reversal in Zone P.9 and an *Asterocyclina* horizon in New Zealand. In the Gippsland and Bass Basins (Fig.1) Partridge (1976a, b) records a dramatic increase in pollen of a *Nothofagus* group in the early Middle Eocene, preceded by local extinctions and followed by impoverished assemblages. Older floras indicated very warm conditions (see also Wopfner et al., 1974) and so did younger (Late Eocene) assemblages, though not so warm as in the Early Eocene. These generalizations agree well with the marine record. The overturn is Gill's (1975) "revolution" in which the Tertiary conifer/*Nothofagus* flora replaced the Mesozoic fern/conifer flora.

Lateritization and the Early Tertiary "African", "Indian", and "Australian" surfaces. King (1962) has summarized the similarities among these continental planations, or erosion cycles, with laterites, all of which succeed Gondwana and post-Gondwana cycles and which are succeeded by middle to late Tertiary cycles, with the ages being decidedly vague. Recently, Dingle and Scrutton (1974) correlated the end of the African cycle with Late Eocene uplift and regression around southern Africa. In Western Australia, extensive laterites are generally regarded as mid-Tertiary in age, on the evidence of lateritized Late Eocene (and older) sediments and non-lateritized Middle Miocene (Johnstone et al., 1973), although there is also good evidence for much younger laterites (Playford et al., 1975). In southern-central Australia the mid-Tertiary is clearly the time of extensive silcrete formation (Wopfner, 1974) whereas the best-developed laterite seems to be of pre-Tertiary to earliest Tertiary age (references cited in Daily et al., 1974) as it is in India (Powell and Conaghan, 1973).

Comment is restricted here to a few observations. In western India, laterite profiles are developed on the (Paleocene) Deccan Traps and are overlain by the sediments of the Middle Eocene transgression; in Cutch and Rajasthan (McGowran, 1977a) the laterite occurs sandwiched between Sequences One and Two, as recognized here. In the St. Vincent Basin in southern Australia, a laterite was recognized (Glaessner and Wade, 1958) in what now seems to be precisely the same position. The sediments at and near the foot of the subsequent transgression, from the Carnarvon Basin to the Otway Basin, are noteworthy for their iron content — as limonite clasts and ferric iron staining as well as in the form of glauconite — and it is possible that at least some of the evidence from Western Australia for lateritization of Late Eocene sediments refers to a remobilization of iron deposited from stripping of an earlier surface. It is suggested that the Early—Middle Eocene decline in temperature will mark quite sharply the end of a major episode of lateritization on the continental fragments of Gondwanaland, and that the Late Paleocene—Early Eocene (i.e. Sequence One) was a time of overlap between the end of its development and the onset of its serious dissection.

The "chert problem". This discussion concentrates on the stratigraphic distribution of the cherts in the oceans and on the southern Australian margin.

The Paleocene oceanic distribution coincides with the widest development of oceanic carbonates in Zone P.4, with the maximum development of Sequence One on the continents and with the onset of major tectonism. In strong contrast, the Early—Middle Eocene oceanic distribution is in the gap between the Sequences, and maximizes isochronously at the foot of the first major temperature decline (Fig.7). In both cases, however, the cherty horizons are in calcareous sections which immediately precede hiatuses in a consistent relationship (Fig.3).

Three younger horizons known in southern Australia are plotted in Fig.7. The oldest (Eucla Basin) is close to the top of the oceanic horizon in the southwest Pacific, and is at about Zone P.14. The second (St. Vincent Basin) is sandwiched between a "warm interval" (on foraminiferal evidence) and a eustatic regression. An enormous amount of silica accumulated locally within the time of Zone P.16; it is not known from coeval sections in the Otway Basin but spongolites in Western Australia (Playford et al., 1975) may be of the same age. The third cherty interval occurs in Early Oligocene carbonates of the St. Vincent and Otway Basins; and that horizon correlates very well indeed with the second major decline in temperature.

Berggren and Hollister (1974) argued that the opening of the Arctic part of the Atlantic Ocean allowed an influx of cold bottom water which caused the formation of both Early—Middle Eocene seismic "Horizon A" (by supplying unusually large amounts of silica) and a marginal hiatus (by strong bottom circulation) (see also Berger and Roth, 1975, fig.6). Ramsay (1971) pointed to a link in time between those cherts and possible glaciation on Antarctica; and Herman (1972) suggested that the genesis of Horizon A and similar rocks required the coincidence of sea-floor spreading (continental breakup) and temperature zonation (i.e. cooling). Ramsay (1974) saw a partitioning of silica and carbonate, Eocene equatorial cherts correlating with a shallow CCD.

Berger's (1974) discussion on stratigraphically ordered oceanic chert occurrences concludes: "At times, such as in the mid-Eocene, flooded shelves may have tied up much carbonate but released the available silica to the deep ocean . . . In addition, flooding of shelves prevents dilution of oceanic volcanogenic sediment with continental debris low in mobile silica. A link to global temperature variations (see T.C. Moore, 1972) in such a model would call for carbonate-rich, silica-poor deep-sea deposits during geocratic times, having cold climates and narrow shelves, with much mixing and upwelling at the shelf edges due to intensified seasonal winds, while silica-rich deep-sea deposits, providing for formation of cherts, would characterize times with warm equable climates and wide carbonate shelf seas with deep-water outflow."

The correlation of the early Middle Eocene (oceanic) and earliest Oligocene (continental) chert horizons with isotopic and biogeographic evidence for cooling is too good to be coincidental and supports, with increased precision, suggestions for a causal link. Those cherts occur in relatively carbonate-rich, rather than carbonate-poor, sections. Although the precise dating of the two Eocene horizons in southern Australia leaves something to be desired (McGowran, 1977b), their position in the composite succession is quite consistent with a model of rapid watermass/climatic fluctuations during the time between the two major declines, the silica accumulating in the colder intervals. As noted already, the distribution of silica in the southwest Pacific Eocene is not well resolved biostratigraphically and could well fit the same pattern. That being so, then the link with temperature occurs in both the oceanic and the neritic horizons. This set of observations would seem to disprove Berger's model (or, more fairly, his groundplan for developing a model), the critical points being that the neritic carbonates and oceanic "Horizon A" cited in his Middle Eocene example can be distinguished historically, and that silica seems to follow carbonate, not fractionate from it.

The Paleocene oceanic horizons fit none of these observations or suggestions other than the chert—carbonate association. This unsatisfactory situation can, however, be turned to some advantage because it supports the notion of a fundamental reorganization of oceanic circulation during the Middle Eocene.

Shelf/basin carbonate fractionation in the Middle and Late Eocene. The oceanic calcareous sections are relatively well developed when neritic carbonates are lacking (early Middle Eocene) and vice versa (late Middle to Late Eocene); the contrast probably is stronger than in the impressionist rendering of Fig.6. The widespread hiatuses above the cherts could, perhaps, be related to the remobilization of silica and attendant diagenetic effects, but the consistence of the stratigraphic patterns encourages the belief that any such effect was not significant.

The CCD in the Pacific and Atlantic Oceans was shallow prior to a drop into the Oligocene (Berger, 1974; Berger and Roth, 1975; Van Andel, 1975;

Van Andel et al., 1975). A composite of CCD curves for the oceans including the Indian (Van Andel, 1975) shows that the Pacific curve is the most dramatic: a rise through the Paleocene and Eocene and a very sharp drop at the Eocene/Oligocene boundary which might be almost 2 km in the equatorial zone (Van Andel et al., 1975). The Atlantic and Indian curves differ somewhat in that they slope downwards through the Eocene before steepening. Backtracking procedures (e.g. Berger and Winterer, 1974; Van Andel et al., 1975) were used to obtain these curves. However, the only clay/carbonate contacts of value for locating the CCD in the Indian Ocean are at Sites 213 and 215, which give Early Eocene CCD depths of 3,600 m and 3,300 m on either side of the Ninetyeast Ridge (Pimm and Sclater, 1974), or of about 3,700 m (Gartner, 1974; Van Andel, 1975). There are no later Eocene contacts to constrain Van Andel's curve.

All the sites which display hiatus within carbonates in the Middle (post-Zone P.11) and Late Eocene (Fig.3) are well above the reconstructed CCD (Pimm and Sclater, 1974; Van Andel, 1975). Since this widespread hiatus matches the neritic carbonate Sequence Two, it seems very likely that a massive shelf/basin fractionation effect (Berger, 1970, 1973) began operating in the Middle Eocene, causing the lysocline to rise sharply, and inviting the prediction that the CCD rose sharply from Zone P.11 onwards before falling into the Oligocene. Thus, the Indian Ocean curve may follow the Pacific more closely than Van Andel has shown. A similarity with the Pacific and contrast with the Atlantic would support Berger's (1970) contrast of the Atlantic ("estuarine"; deep-water outflow) to Pacific and Indian ("lagoonal"; deep-water inflow). Finally, the accumulation of oceanic carbonate is more pronounced in Zones P.14—P.15 than immediately before or after (Fig.3). The continental/oceanic contrast in the Zones P.11—P.12 interval (Fig.6) shows a rather rapid response in shelf/basin fractionation, and so one might suspect widespread hiatus in neritic carbonates across the Middle/Late Eocene boundary. The absence of a clearly recognized hiatus disrupting Sequence One (Fig.5) is merely a reflection of inadequate biostratigraphic resolution in that facies.

Summary: significance of Australia/Antarctica separation. Kennett and his co-workers have discussed extensively the critical position of the Eocene/Oligocene boundary and the Oligocene in Cainozoic history: the development of circum-Antarctic circulation and lessening of oceanic meridional circulation, the decline in temperature to the point where sea ice is produced extensively in the Antarctic, and the development of very cold Antarctic bottom water, modern thermo-haline circulation and bottom waters of the oceans (the "psychrosphere"), are all considered to happen at that time (Kennett et al., 1975; Shackleton and Kennett, 1975; Kennett and Shackleton, 1976). An important part of that history is the suggestion that circum-Antarctic sea-floor spreading did not allow significant deep-water circulation to the south of Australia and South America until the Oligocene (see also Pimm and Sclater, 1974; Edwards, 1975; Deighton et al., 1976).

But Australia/Antarctica separation seems to have been more influential in the Middle Eocene than is generally acknowledged.

Temperatures were relatively high in the Paleocene and Early Eocene, and yet larger foraminifera did not colonize the neritic carbonate environments available in the south. In Zone P.4 there is not the negative correlation perceived later between oceanic and neritic carbonates, nor between chert horizons and temperature. The patterns and correlations of diverse evidence in the Middle and Late Eocene are lacking before then. An important reorganization of circulation and climate (using both terms very broadly) must have occurred, and the Early/Middle Eocene transition is the logical interval to scrutinize. The correlation of oceanic subsidence, termination of continental Sequence One including marine regression, and climatic deterioration invites speculation that "passive tectonism" stimulated the climatic change through regression and increased continentality.

The allochronous top horizon of the oceanic cherts in the Eocene — with the Eucla Basin in southern Australia belonging in this respect with the southwest Pacific — indicates significant deflection of southern ocean waters to the south of Australia for the first time. This deflection led to a significant quarantining of southern waters so that by Zone P.13 time the widespread migrations of tropical-type organisms could take place. Abundant evidence for relatively warm conditions in southern Australia and New Zealand at several horizons during the later Middle and Late Eocene is not matched by a restoration of isotopic temperatures to previous levels on the South Campbell Plateau. There may have been a decoupling of watermass temperatures between high and lower latitudes similar to the Neogene decoupling observed by Savin et al. (1975). (However, intra-Eocene climatic changes were bipolar. Steineck (1971) cites planktonic foraminiferal changes across the Early/Middle Eocene boundary as evidence for early Middle Eocene "refrigeration" in California, whereas Adams (1973) notes the importance of the larger foraminifer *Discocyclina* in the (probably later Middle) Eocene of Alaska.) The onset of shelf/basin carbonate fractionation during the Middle Eocene supports the other evidence for significant circum-Antarctic circulation by then.

The evidence for ice-rafting of sediments into the southeast Pacific (Margolis and Kennett, 1971) during the Eocene is taken now to indicate only "any Antarctic glaciation restricted to higher elevations" (Kennett et al., 1975). The planktonic foraminiferal evidence for ages (Margolis and Kennett, 1971; Jenkins, 1974) give a range for the two possible episodes of Zones P.6—P.7 and Zones P.11—P.14 respectively. The earlier range shows no correlation with events discussed here. The later range could, perhaps, match lows in Zones P.11—P.12 or Zone P.14. A restricted glaciation in the time of Zones P.11—P.12 is the only possibility at all convincing (Fig.7); it would interpolate between the warmer times of Zones P.9 and P.13; and the lack of a well-dated geological record in that interval would explain the lack of any evidence in southern Australia of Antarctic glaciation (McGowran, 1973b). In any case, glacio-eustacy will be greatly subordinate to tectono-eustacy in explaining the continental stratigraphic record.

CONCLUDING REMARKS

Episodic history

The literature on the Indian Ocean region and Australia contains many references to diachronous patterns, be they in facies changes in oceanic sediments, transgressions on the continents, climatic changes, tectonism and geosynclinal evolution, or in the stratigraphic distribution of fossils. The present review gives a different impression; rather than *diachronous* geohistory and biohistory one acquires instead a sense of rapid or virtually *isochronous* events, and where offsets in time are detected the pattern accordingly is *allochronous*. And this very broad generalization recalls the more specific question of episodicity or continuity of orogenic movements which, as e.g. Trümpy (1973) has summarized, is an opposition of (1) relatively short movements as phases globally contemporaneous, and separated by longer times of quiescence (Hans Stille), to (2) continuous movements over a long time with no discrete phases other than sampling effects (James Gilluly).

Pimm et al. (1974) showed that diachronous facies changes on the Ninetyeast Ridge were thoroughly consistent with the notion of rapid northward movement and concomitant rapid sinking. This was good confirmation from a ridge of the pattern of ocean basin sedimentation generally, whereby diachronous facies changes are generated as different sites on a moving plate move through the same depth and productivity zones at different times (Heezen et al., 1973; Hesse et al., 1974). Although there is no serious questioning of this model, which is the basis for plate stratigraphy (Berger and Winterer, 1974), it is incomplete in lacking the isochronous overprint on the oceanic stratigraphic record.

Hermes (1974) detected no episodes in the orogenic development of Irian Jaya; as suggested above, his data do not demand that conclusion for the Early Tertiary.

Jenkins (1974) has stressed the diachronous distribution of initial appearances and extinctions among planktonic foraminifera in the southern extratropical (Austral) region and between the Austral and the Tropical—subtropical Belts (for a contrary view, suggesting isochronous biostratigraphic events including non-phyletic and disjunct distributions, see McGowran, 1977b). Jenkins sees a large-scale pattern of diachronous climatic deterioration beginning in the Antarctica in the Late Eocene and reaching the tropics in the late Early Oligocene; Kennett et al. (1975) and Kennett and Shackleton (1976) discuss his "time sequential paleobiogeographic model" further. In continental stratigraphic sections, diachronism was stressed as an outcome of Eocene correlations by D.J. Taylor (1971) and McGowran (1971, 1973a, b) and adopted by Kennett et al. (1975) and Deighton et al. (1976). The same theme is evident in suggestions relating continental "drift" to sedimentation and biogeography: the forest cover changed as Australia was rafted north to warmer climates (Raven and Axelrod, 1972); the "abrupt" change to carbonate sedimentation in the Great Australian Bight

Basin occurred possibly for the same reason (Willcox, 1976) or even, merely, when access to truly oceanic conditions was first possible (Deighton et al., 1976).

Thus, diachronism, stressed from the earliest stages of one's exposure to stratigraphy, the facies concept and basin analysis, and to the stern injunction to beware the traps of lithological correlation and the temptations of an incomplete geological record, pervades historical thought. At one level (for example, interfingering lithologies in basin stratigraphy) there can be no doubt about diachronism and its importance. At another level it is easy enough to develop a circular argument, and to perceive episodic synchronism where one seeks it; indeed, the Stillean approach tended to derive a historical sequence from prediction of orogenic timing, rather than vice versa (e.g. Rutten, 1969). But with careful attention to chronology — and to the quality of the evidence! — Trümpy (1973) has come to see episodic synchronism in Alpine tectonic evolution: "when the author set out to gather information on the timing of orogenic events, he started as a convinced Gillulyan; to his own surprise, he has ended up as a moderate Stillean". The present review, which has little else in common with Trümpy's, comes to the same conclusion. It provides some justification for promoting Trümpy's "neo-Stillean" *Weltanschauung.*

"Horizontal" versus "vertical" explanation

The picture of isochronous events overriding local geological environments suggests that "local" explanations must fall short. Examples include: post-collision downbuckling of the Indian margin to explain the Middle Eocene transgression (Powell and Conaghan, 1973); regional explanation of the hiatus at the Cretaceous/Tertiary boundary in Western Australia (Veevers and Johnstone, 1974); and the initiation of carbonate sedimentation in southern Australia, as mentioned above. But local explanation is most comprehensive in the recent models for the evolution of passive continental margins (Falvey, 1974; Kinsman, 1975; Deighton et al., 1976) in which a sequence of events and their timing are determined by subcrustal processes.

In Falvey's classification and terminology, the southern margin of Australia records the following lithotectonic succession: basement (Palaeozoic); pre-(proto) rift intracratonic basin sequence (Late Jurassic—Early Cretaceous); rift onset unconformity; rift valley sequence (Late Cretaceous); break-up unconformity (Cretaceous/Tertiary boundary); post-rift-valley stage onlap (Sequence One, herein); post-breakup prograded section (Sequence Two, herein, and the remainder of the Cainozoic). Thermal changes in the lithosphere caused two cycles of uplift/erosion/subsidence. Falvey (1974) discusses variation in which one or more stages may be missing, but the theme is crustal response to subcrustal processes. Deighton et al. (1976) have modified the succession, partly after a review by Boeuf and Doust (1975), and the breakup unconformity is now between Sequences One and Two (herein) with Sequence One identified as an infra-breakup phase. There is much greater commitment to stratigraphic detail in this version, particularly in their palaeogeographic maps (spanning Paleocene—Late Oligocene) in which, however, the distinction between facies observed and dated, and facies generated by hypothetical circulation patterns based on plate-tectonic reconstruction, is not at all clear. Their facies map for the early Middle Eocene is sheer fiction; it shows sedimentation across 45° longitude of continental margin for a time interval represented by not one accurately dated sample known to this writer, and warm conditions are postulated in defiance of all evidence. But the question of timing is more fundamental (Fig.9). The "infra-breakup phase" is Sequence One, more widespread than a model for a single passive margin can explain; the "breakup unconformity" is found on two continents on separate plates; the onset of bryozoan shelf carbonate facies, explained as due to sufficient subsidence of the margin with continued spreading into the later Middle Eocene, is also intercontinental and isochronous on two plates. Rather than a subcrustal cycle generating a timed succession of events in the formation of a continental margin, it seems more likely that each event (phase) in a homotaxial succession is a local response to platewide or global tectonic rearrangement. For example, on the evidence available for correlation it seems possible that India/Asia collision, Australia/Antarctica separation, the change from Tasman Sea to Coral Sea spreading, collision north of Australia and the oceanic, mid-plate unconformity are all part of the same geodynamic climax, timed by no more than the geographic situation of continents relative to converging plate boundaries. To that extent the timing of tectonic rearrangement is random although its configuration is not: the configuration of Australia/Antarctica separation may have been predetermined as long ago as the Permian (McGowran, 1973b; rejected

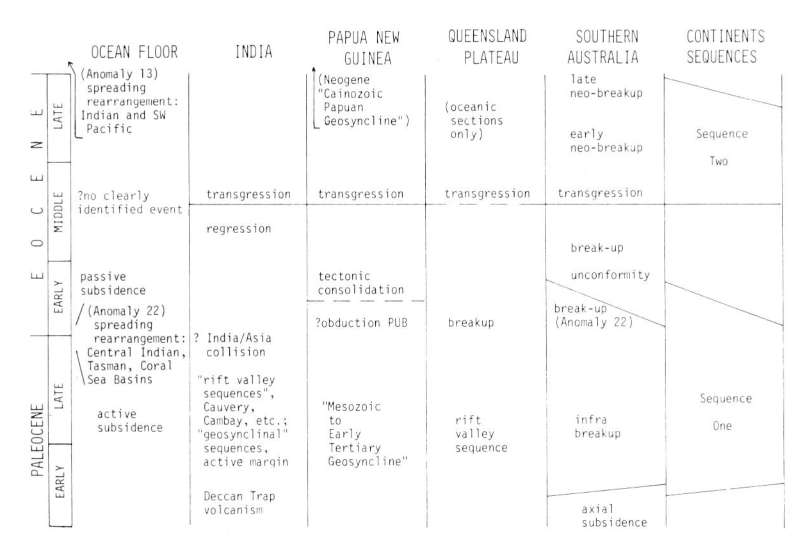

Fig.9. Correlation of diverse events, intended to demonstrate that the stratotectonic succession on the new, passive margin of southern Australia (Deighton et al., 1976) should not be analysed in isolation, as discussed in text. *PUB* = Papuan Ultramafic Belt.

by Deighton et al., 1976 because it does not fit their lithotectonic/cyclic model). On several occasions Kent (e.g. 1975) has made the same point. Similarly, the likelihood that the Coral Sea—Queensland Plateau region evolved isochronously with southern Australia in at least two respects (initial emplacement of oceanic crust; onset of marginal Sequence Two) is seen as more important than is resolution of dissension over the existence of a rift valley stage (L.W.H. Taylor, 1975; Mutter, 1975; Veevers and McElhinny, 1976).

A note on the Haug Effect

The Haug Effect states, in summary, that "episodes of sea-floor spreading, times of growing oceanic rise systems, eustatic sea-level rise, and orogenic cycles are all coincident" (J.G. Johnson, 1974). Much of the discussion on the perception of such timing arises from confusion over scale of time and precision in correlation and also over what is meant by orogeny, but Rona (1973, 1974) identifies the distinction between orogeny plus stratigraphic sequence on the one continent, on the one hand, and the global response to eustatic phenomena on the other.

For several reasons the present study does not clearly support the Haug Effect. The basic notion in Johnson's model is that geosynclinal orogeny and marginal hiatus correlate with marine deposition in an adjacent cratonic sequence, whereas the reverse is true here: hiatus and sequence on active margins are in phase, not out of phase, with hiatus and sequence on the passive margins. Secondly, the stratigraphic patterns are intercontinental and Rona's (1973, 1974) emphasis on global patterns is supported accordingly. Finally, there is a difference in time scale of an order of magnitude; in Johnson's model, three sequences span 200 or 300 m.y. However, there is sufficient coherence in the Early Tertiary history outlined here to justify further and more sophisticated (especially chronological) scrutiny of the Haug Effect.

SUMMARY

Foraminiferal correlations and biofacies changes bear on the timing and interrelationships of tectonic and oceanic/climatic events. There are two stratigraphic *sequences* on the continents (Pakistan—India, Australian region).

Sequence One (Paleocene—Early Eocene) begins in Zones P.1—P.3 in shelf carbonates or terrigenous sediments on the passive margins and in shallow- to deeper-water facies on the active margins, and terminates with regression, hiatus, or biostratigraphic non-identification in or close to Zone P.7. On the Mascarene Plateau, Chagos—Laccadive Ridge and Ninetyeast Ridge, successional foraminiferal assemblages record rapid and concerted subsidence during the later Paleocene which was interrupted by an isochronous hiatus across the Paleocene/Eocene boundary. The resumption of accumulation is paralleled by a change from calcareous ooze to clay in the deep oceanic basins and by the onset of marginal regression, i.e. termination of Sequence One.

Early in the Middle Eocene (Zone P.11; ca. 47—48 Ma) the top of the chert horizon in pelagic carbonates is virtually isochronous in the Indian Ocean when there is oceanic foraminiferal evidence for a "cool interval". Later in the Middle Eocene (Zones P.12—P.13; ca. 45—46 Ma) Sequence Two begins isochronously around the margins as a transgression. Middle and Late Eocene shelf carbonates must have been voluminous before Oligocene erosion, but oceanic foraminiferal assemblages are poorly preserved, or missing. Pantropical, neritic, larger foraminifera were on New Guinea, India, the Chagos—Laccadive Ridge and Madagascar in the Maastrichtian and the Paleocene, but have not been found in Western Australia or on the Ninetyeast Ridge. In contrast, *Discocyclina* migrated southward to both places at ca. 45 Ma, and planktonic assemblages (Zone P.13) also indicate a temporary expansion of the tropical belt.

The 56—52 Ma oceanic hiatus is coeval with the first India/Eurasia collision, an orogeny in Timor, (?) emplacement of the Papuan Ultramafic Belt, opening of the Coral Sea Basin, termination of spreading in the Tasman Sea, Australia/Antarctica separation and tectonic rearrangement in the central Indian Ocean. The immediately subsequent marginal regression and hiatus are related to resumption of subsidence in the Indian Ocean when India's northward motion and sea-floor spreading rates both declined sharply.

That regression was coeval with a terrestrial floral change in Australia and with the beginning of a temperature decline. The early Middle Eocene chert horizon at the foot of that decline records the last major oceanic event in the Indian Ocean before significant deflection of oceanic water to the south of Australia; the chert horizons are allochronously younger in the southwest Pacific Ocean. The subsequent Middle Eocene transgression points to a tectono-eustatic event at about Anomaly 20, 46—47 Ma. The transgression plus quarantining of Antarctic water stimulated evolutionary radiation among and excursions southward by tropical foraminifera. Neritic carbonate accumulations probably caused the calcite compensation depth to rise in response to shelf/basin fractionation; this effect was less marked or absent during the Maastrichtian and Paleocene. There was a significant advance in the evolution of the present oceanic/climatic regime during the Middle Eocene.

Thus, diverse geohistorical events occurred episodically but in concert across the Indian and Australian, then Indo-Australian Plate. Isochronism is seen between continent and ocean; between oceanic carbonates, neritic carbonates and terrigenous clastics; and between active margins, passive margins and continental platforms. Essentially local explanations advanced recently for phenomena perceived locally must fall short when the history is platewide. Stratotectonic events evidently are more likely to provide world-wide tectonic markers than are the traditional morphotectonic folding episodes.

ADDENDUM

Developments since the correlation charts were prepared include the following items:

Hardenbol and Berggren (1976) have moved the Middle/Late Eocene boundary upward because the traditionally Late Eocene Bartonian Stage is now known to contain Middle Eocene-type planktonic microfossils. The boundaries are now: Eocene/Oligocene, 37 Ma; Middle/Late Eocene, 40 Ma; older boundaries, unchanged. The revision means that Zones P.10—P.14 are expanded and Zones P.15—P.16 are compressed. This is intuitively acceptable because the relative lack of biostratigraphic refinement in Late Eocene pelagic and hemipelagic sections at low latitudes has never been explained adequately, and because the compressing of foraminiferal-biostratigraphic (non-phyletic) and palaeoclimatic events in southern Australasia heightens the sense of rapid oscillation before the major change between the Eocene and the Oligocene. Otherwise, nothing based on the correlations discussed here is changed significantly.

Smith and Davies (1976) describe deep-water sediments and volcanics in southeast Papua and Belford (1976) presents evidence — abundant planktonic foraminifera in random thin-sections — for their Middle Eocene age. This suggests a record in Sequence Two of an intermediate step in the shift of the locus of volcanic activity after the Early Eocene (McKenzie, 1975; cited above).

B.D. Johnson et al. (1976) have presented the most detailed spreading history so far for the eastern Indian Ocean. In adopting a four-stage instead of three-stage synopsis (Veevers and McElhinny, 1976; cited above) they highlight the importance of the event at 53 Ma (Anomaly 22) when the spreading ridge west of the Ninetyeast Ridge jumped to the south, linking up with the new ridge between Australia and Antarctica.

REFERENCES

Adams, C.G., 1970. A reconsideration of the East Indian Letter classification of the Tertiary. Bull. Br. Mus. (Nat. Hist.), Geol., 19: 85—137.

Adams, C.G., 1973. Some Tertiary foraminifera. In: A. Hallam (Editor), Atlas of Palaeobiogeography. Elsevier, Amsterdam, pp.453—468.

Andrews, J.E., Packham, G.H., Eade, J.V., Holdsworth, B.K., Jones, D.L., Klein, G. DeV., Kroenke, L.W., Saito, T., Shafik, S., Stoeser, D.B. and Van der Lingen, G.J., 1975. Initial Reports of the Deep Sea Drilling Project, 30. U.S. Govt. Printing Office, Washington, D.C.

Anonymous, 1974. Geology of Western Australia. West. Aust. Geol. Surv. Mem., 2: 541 pp.

Audley-Charles, M.G. and Carter, D.J., 1972. Palaeogeographical significance of some aspects of Palaeogene and early Neogene stratigraphy and tectonics of the Timor Sea region. Palaeogeogr., Palaeoclimatol., Palaeoecol., 11: 247—264.

Belford, D.J., 1976. Appendix: Foraminifera and age of samples from south-eastern Papua. In: I.H. Smith and H.L. Davies (Editors), Geology of the South-eastern Papuan Mainland. Bull. Bur. Miner. Resour., Geol. Geophys., Aust., 165: 73—82.

Berger, W.H., 1970. Biogenous deep-sea sediments: fractionation by deep-sea circulation. Geol. Soc. Am., Bull., 81: 1385—1402.

Berger, W.H., 1973. Cenozoic sedimentation in the eastern tropical Pacific. Geol. Soc. Am., Bull., 84: 1941—1954.

Berger, W.H., 1974. Deep-Sea sedimentation. In: C.A. Burk and C.L. Drake (Editors), The Geology of Continental Margins. Springer, New York, N.Y., pp.203—241.

Berger, W.H. and Roth, P.H., 1975. Oceanic micropaleontology: Progress and prospect. Rev. Geophys. Space Phys., 13: 561—585.

Berger, W.H. and Winterer, E.L., 1974. Plate stratigraphy and the fluctuating carbonate line. In: K.J. Hsü and H. Jenkyns (Editors), Pelagic Sediments on Land and under the Sea. Int. Assoc. Sedimentol., Spec. Publ., 1: 11—48.

Berggren, W.A., 1972. A Cenozoic time-scale — some implications for regional geology and paleobiogeography. Lethaia, 5: 195—215.
Berggren, W.A. and Hollister, C.D., 1974. Paleogeography, paleobiogeography and the history of circulation in the Atlantic Ocean. In: W.W. Hay (Editor), Studies in Paleo-Oceanography. Soc. Econ. Paleontol. Mineral., Spec. Publ., 20: 126—186.
Boeuf, M.G. and Doust, H., 1975. Structure and development of the southern margin of Australia. J. Aust. Pet. Explor. Assoc., 15: 33—43.
Brown, C.M., Pieters, P.E. and Robinson, G.P., 1975. Stratigraphic and structural development of the Aure Trough and adjacent shelf and slope areas. J. Aust. Pet. Explor. Assoc., 15: 61—71.
Burk, C.A. and Drake, C.L. (Editors), 1974. The Geology of Continental Margins. Springer, New York, N.Y.
Carter, D.J., Audley-Charles, M.G. and Barber, A.J., 1976. Stratigraphic analysis of island arc—continental margin collision in eastern Indonesia. J. Geol. Soc., 132: 197—198.
Colbert, E.H., 1973. Wandering Lands and Animals. Dutton, New York, N.Y., 323 pp.
Coleman, P.J. and Packham, G.H., 1976. The Melanesian Borderlands and India—Pacific plates' boundary. In: M.F. Glaessner (Editor), The Geosciences in Australia. Earth-Sci. Rev., 12: 197—233.
Crawford, A.R., 1974. The Indus suture line, the Himalaya, Tibet and Gondwanaland. Geol. Mag., 111:369—383.
Curray, J.R. and Moore, D.G., 1974. Sedimentary and tectonic processes in the Bengal Deep-Sea Fan and Geosyncline. In: C.A. Burk and C.L. Drake (Editors). The Geology of Continental Margins. Springer, New York, N.Y., pp.617—627.
Daily, B., Twidale, C.R. and Milnes, A.R., 1974. The age of the lateritized surface on Kangaroo Island and adjacent areas of South Australia. J. Geol. Soc. Aust., 21: 387—392.
Davies, H.L., 1976. Papua New Guinea ophiolites. Int. Geol. Congr., 25th, Excursion Guide No. 52A: 13 pp.
Davies, H.L. and Smith, I.E., 1971. Geology of eastern Papua. Geol. Soc. Am., Bull., 82: 3299—3312.
Davies, T.A., Weser, O.E., Luyendyk, B.P. and Kidd, R.B., 1975. Unconformities in the sediments of the Indian Ocean. Nature, 253: 15—19.
Deighton, I., Falvey, D.A. and Taylor, D.J., 1976. Depositional environments and geotectonic framework: southern Australian continental margin. J. Aust. Pet. Explor. Assoc., 16: 25—26.
Dingle, R.V. and Scrutton, R.A., 1974. Continental breakup and the development of post-Paleozoic sedimentary basins around southern Africa. Geol. Soc. Am., Bull., 85: 1467—1474.
Douglas, J.G. and Ferguson, J.A. (Editors), 1976. Geology of Victoria. Geol. Soc. Aust., Spec. Publ., 5: 528 pp.
Dow, D.B., 1975. Geology of Papua New Guinea. In: C.L. Knight (Editor), Economic Geology of Australia and Papua New Guinea, 1. Metals. Australas. Inst. Min. Metall., Monogr., 5: 823—836.
Edwards, A.R., 1975. Southwest Pacific Cenozoic paleogeography and an integrated Neogene paleocirculation model. Initial Reports of the Deep Sea Drilling Project, 30. U.S. Govt. Printing Office, Washington, D.C., pp.667—684.
Falvey, D.A., 1974. The development of continental margins on plate tectonic theory. J. Aust. Pet. Explor. Assoc., 14: 95—106.
Fischer, A.G., 1974. Origin and growth of basins. In: A.G. Fischer and S. Judson (Editors), Petroleum and Global Tectonics. Princeton University Press, Princeton, N.J., pp.47—79.
Flemming, N.C. and Roberts, D.G., 1973. Tectono-eustatic changes in sea level and seafloor spreading. Nature, 243: 19—22.
Gartner, S., 1974. Nannofossil biostratigraphy, Leg 22, Deep Sea Drilling Project. Initial Reports of the Deep Sea Drilling Project, 22. U.S. Govt. Printing Office, Washington, D.C., pp.577—599.

Glaessner, M.F. and Wade, M., 1958. The St. Vincent Basin. In: M.F. Glaessner and L.W. Parkin (Editors), Geology of South Australia. J. Geol. Soc. Aust., 5 (2): 115—126.
Hardenbol, J. and Berggren, W.A., 1976. A new Paleogene numerical time scale. Int. Geol. Congr., 25th, Symp. 106.6, Abstr., 2 pp. (unpublished).
Harris, W.K., 1971. Tertiary stratigraphic palynology, Otway Basin. In: H. Wopfner and J.G. Douglas (Editors), The Otway Basin of southeastern Australia. Geol. Survs. South Aust. Vict., Spec. Bull., pp.67—87.
Hayes, D.E. and Ringis, J., 1973. Sea floor spreading in the Tasman Sea. Nature, 243: 454—458.
Hays, J.D. and Pitman, W.C., 1973. Lithospheric plate motion, sea level changes and climatic and ecological consequences. Nature, 246:18—22.
Heezen, B.C., MacGregor, I.D., Foreman, H.P., Forristall, G., Hekel, H., Hesse, R., Hoskins, R.H., Jones, E.J.W., Krasheninnikov, V.A., Okada, H. and Ruef, M.H., 1973. Diachronous deposits: a kinematic interpretation of the post-Jurassic sedimentary sequence on the Pacific Plate. Nature, 241: 25—32.
Heirtzler, J.R., Dickson, G.O., Herron, E.M., Pitman, W.C. and Le Pichon, X., 1968. Marine magnetic anomalies, geomagnetic field reversals, and motions of the ocean floor and continents. J. Geophys. Res., 73: 2119—2136.
Herman, Y., 1972. Origin of deep sea cherts in the North Atlantic. Nature, 238: 392—393.
Hermes, J.J., 1974. West Irian. In: A.M. Spencer (Editor), Mesozoic—Cenozoic Orogenic Belts: Data for Orogenic Studies. Geol. Soc., Spec. Publ., 4: 475—490.
Hesse, R., Foreman, H.P., Forristall, G.Z., Heezen, B.C., Hekel, H., Hoskins, R.H., Jones, E.J.W., Kaneps, A.G., Krasheninnikov, V., MacGregor, I. and Okada, H., 1974. Walther's facies rule in pelagic realm — a large-scale example from the Mesozoic—Cenozoic Pacific. Z. Dtsch. Geol. Ges., 125: 151—172.
Jenkins, D.G., 1974. Paleogene planktonic foraminifera of New Zealand and the Austral region. J. Foraminiferal Res., 4: 155—170.
Jenkins, D.G., 1975. Cenozoic planktonic foraminiferal biostratigraphy of the south-western Pacific and Tasman Sea — DSDP Leg 22. Initial Reports of the Deep Sea Drilling Project, 29. U.S. Govt. Printing Office, Washington, D.C., pp.449—467.
Johnson, B.D., Powell, C. McA. and Veevers, J.J., 1976. Spreading history of the eastern Indian Ocean and Greater India's flight from Antarctica and Australia. Geol. Soc. Am., Bull., 87: 1560—1566.
Johnson, J.G., 1971. Timing and coordination of orogenic, epeirogenic, and eustatic events. Geol. Soc. Am., Bull., 82: 3263—3298.
Johnson, J.G., 1974. Sea-floor spreading and orogeny: correlation or anticorrelation? Geology, 2: 199—201.
Johnstone, M.H., Lowry, D.C. and Quilty, P.G., 1973. The geology of southwestern Australia — a review. J. R. Soc. West. Aust., 56: 5—15.
Karunakaran, C., Ray, K.K. and Saha, S.S., 1976. Implication of the mobile belt of the Indies Archipelago on the concept of continental drift between India and Australia. J. Geol. Soc. India, 17: 309—321.
Kennett, J.P. and Shackleton, N.J., 1976, Oxygen isotopic evidence for the development of the psychrosphere 38 Myr ago. Nature, 260: 513—515.
Kennett, J.P., Houtz, R.E., Andrews, P.B., Edwards, A.R., Gostin, V.A., Hajos, M., Hampton, M., Jenkins, D.G., Margolis, S.V., Ovenshine, A.T. and Perch-Nielsen, K., 1975. Cenozoic paleoceanography in the southwest Pacific Ocean, Antarctic glaciation, and the development of the circum-Antarctic current. Initial Reports of the Deep Sea Drilling Project, 29. U.S. Govt. Printing Office, Washington, D.C., pp.1155—1169.
Kent, P.E., 1975. Mesozoic development of aseismic continental shelves. In: Geodynamics today: A review of the earth's dynamic processes. R. Soc., pp.119—121.
King, L.C., 1962. Morphology of the Earth. Oliver and Boyd, Edinburgh, 699 pp.
Kinsman, D.J.J., 1975. Rift valley basins and sedimentary history of trailing continental margins. In: A.G. Fischer and S. Judson (Editors), Petroleum and Global Tectonics. Princeton University Press, Princeton, N.J., pp. 83—126.

Lange, R.T., 1970. The Maslin Bay flora, South Australia. Neues Jahrb. Geol. Paläontol., Monatsh., 8: 486—490.
Leslie, R.B., Evans, H.J. and Knight, C.L. (Editors), 1976. Economic Geology of Australia and Papua New Guinea, 3. Petroleum. Australas. Inst. Min. Metall., Monogr., 7: 541 pp.
Margolis, S.V. and Kennett, J.P., 1971. Cenozoic paleoglacial history of Antarctica recorded in subantarctic deep-sea cores. Am. J. Sci., 271: 1—36.
Martin, P.G., 1975. Marsupial biogeography in relation to continental drift. Mém. Mus. Nat. Hist. Nat. Sér. A, Zool., 88: 216—237.
McGowran, B., 1968. Late Cretaceous and Early Tertiary correlations in the Indo-Pacific region. Geol. Soc. India, Mem., 2: 335—360.
McGowran, B., 1971. Australia—Antarctica separation and the Eocene transgression in southern Australia. S. Aust. Dep. Min., Rep. 71/78 (unpublished).
McGowran, B., 1973a. Observation Bore No. 2, Gambier Embayment of the Otway Basin: Tertiary micropalaeontology and stratigraphy. S. Aust. Dep. Min. Miner. Resour. Rev., 135: 43—55.
McGowran, B., 1973b. Rifting and drift of Australia and the migration of mammals. Science, 180: 759—761.
McGowran, B., 1974. Foraminifera. Initial Reports of the Deep Sea Drilling Project, 22. U.S. Govt. Printing Office, Washington, D.C., pp.609—628.
McGowran, B., 1977a. Maastrichtian to Eocene foraminiferal assemblages in the northern and eastern Indian Ocean region: correlations and historical patterns. In: Indian Ocean Geology and Biostratigraphy. American Geophysical Union, in press.
McGowran, B., 1977b. Early Tertiary foraminiferal biostratigraphy in southern Australia: progress report. Bull. Bur. Miner. Resour., Geol. Geophys., Aust., in press.
McGowran, B. and Wade, M., 1967. Latitudinal variation in the foraminiferal content of biogenic sediments. Aust. N.Z. Assoc. Advancement of Science, 39th Congress, Abstracts Section C (Geology): A9—A11.
McKenzie, D.E., 1975. Volcanic and plate tectonic evolution of central Papua New Guinea. In: D.A. Falvey and G.H. Packham (Editors), Southwest Pacific Workshop. Aust. Soc. Explor. Geophys., Bull., 6 (2/3): 66—68.
Moore, D.G., Curray, J.R., Raitt, R.W. and Emmel, F.J., 1974. Stratigraphic-seismic section correlations and implications to Bengal Fan history. Initial Reports of the Deep Sea Drilling Project, 22. U.S. Govt. Printing Office, Washington, D.C., pp.403—412.
Moore, T.C., 1972. DSDP: Successes, failures, proposals. Geotimes, 17 (7): 29—31.
Mutter, J.C., 1975. Basin evolution and marginal plateau subsidence in the Coral Sea. In: D.A. Falvey and G.H. Packham (Editors), Southwest Pacific Workshop. Aust. Soc. Explor. Geophys., Bull., 6 (2/3): 35—37.
Nagappa, Y., 1959. Foraminiferal biostratigraphy of the Cretaceous—Eocene succession in the India—Pakistan—Burma region. Micropaleontology, 5: 145—192.
Partridge, A., 1976a. The geological expression of eustacy in the Early Tertiary of the Gippsland Basin. J. Aust. Pet. Explor. Assoc., 16: 73—79.
Partridge, A., 1976b. The palaeoclimatic control on southern Australian Tertiary spore-pollen assemblages. Int. Geol. Congr., 25th, Abstr., 1: 331—332.
Pimm, A.C. and Sclater, J.G., 1974. Early Tertiary hiatuses in the northeastern Indian Ocean. Nature, 252: 362—365.
Pimm, A.C., McGowran, B. and Gartner, S., 1974. Early sinking history of Ninetyeast Ridge, north-eastern Indian Ocean. Geol. Soc. Am., Bull., 85: 1219—1224.
Playford, P.E., Cope, R.N., Cockbain, A.E., Low, G.H. and Lowry, D.C., 1975. Phanerozoic. In: Geology of Western Australia. West. Aust. Geol. Surv., Mem., 2: 223—433.
Powell, C.McA. and Conaghan, P.J., 1973. Plate tectonics and the Himalayas. Earth Planet. Sci. Lett., 20: 1—12.
Powell, C.McA. and Conaghan, P.J., 1975. Tectonic models of the Tibetan Plateau. Geology, 3: 727—731.
Ramsay, A.T.S., 1971. The investigation of Lower Tertiary sediments from the North

Atlantic. In: A. Farinacci (Editor), Proceedings of the II Planktonic Conference, Roma 1970. Edizioni Tecnoscienza, Roma, pp.1039—1055.
Ramsay, A.T.S., 1974. The distribution of calcium carbonate in deep sea sediments. In: W.W.Hay (Editor), Studies in Paleo-oceanography. Soc. Econ. Paleontol. Mineral., Spec. Pap., 20: 58—76.
Raven, P.H. and Axelrod, D.I., 1972. Plate tectonics and Australasian paleobiogeography. Science, 176: 1379—1386.
Richards, J.R., Cooper, J.A., Webb, A.W. and Coleman, P.J., 1966. Potassium—argon measurements of the age of basal schists in the British Solomon Islands. Nature, 211: 1251—1252.
Rona, P.A., 1973. Relations between rates of sediment accumulation on continental shelves, sea-floor spreading, and eustacy inferred from the central North Atlantic. Geol. Soc. Am., Bull., 84: 2851—2872.
Rona, P.A., 1974. Relation in time between eustacy and orogeny: a presently indeterminate problem. Geology, 2: 201—202.
Rutten, M.G., 1969. The Geology of Western Europe. Elsevier, Amsterdam, 520 pp.
Ryburn, R.J., 1976. The Median Tectonic Line in New Guinea — a continent—island arc collision suture. BMR Symp., 5th, Abstr., BMR J. Aust. Geol. Geophys., 1: 257.
Sahni, A. and Kumar, V., 1974. Palaeogene paleobiogeography of the Indian subcontinent Palaeogeogr., Palaeoclimatol., Palaeoecol., 15: 209—226.
Savin, S.M., Douglas, R.G. and Stehli, F.G., 1975. Tertiary marine paleotemperatures. Geol. Soc. Am., Bull., 86: 1499—1510.
Sclater, J.G. and Fisher, R.L., 1974. Evolution of the east central Indian Ocean, with emphasis on the tectonic setting of the Ninetyeast Ridge. Geol. Soc. Am., Bull., 85: 683—702.
Sclater, J.G., Jarrard, R., McGowran, B. and Gartner, S., 1974. Comparison of the magnetic and biostratigraphic time scales since the Late Cretaceous. Initial Reports of the Deep Sea Drilling Project, 22. U.S. Govt. Printing Office, Washington, D.C., pp.381—386.
Sclater, J.G., Luyendyk, B.P. and Meinke, L., 1976. Magnetic lineations in the southern part of the Central Indian Basin. Geol. Soc. Am., Bull., 87: 371—378.
Shackleton, N.J. and Kennett, J.P., 1975. Paleotemperature history of the Cenozoic and the initiation of Antarctic glaciation: oxygen and carbon isotope analyses in DSDP Sites 277, 279, and 281. Initial Reports of the Deep Sea Drilling Project, 29. U.S. Govt. Printing Office, Washington, D.C., pp.743—755.
Sloss, L.L., 1963. Sequences in the cratonic interior of North America. Geol. Soc. Am., Bull., 74: 93—114.
Smith, I.E. and Davies, H.L., 1976. Geology of the southeast Papuan mainland. Bull. Bur. Miner. Resour., Geol. Geophys., Aust., 165: 86 pp.
Spencer, A.M. (Editor), 1974. Mesozoic—Cenozoic Orogenic Belts: Data for Orogenic Studies. Geol. Soc., Spec. Publ., 4: 809 pp.
Steineck, P.L., 1971. Middle Eocene refrigeration: new evidence from California planktonic foraminiferal assemblages. Lethaia, 4: 125—129.
Stonely, R., 1974. Evolution of the continental margins bordering a former southern Tethys. In: C.A. Burk and C.L. Drake (Editors), The Geology of Continental Margins. Springer, New York, N.Y., pp.889—903.
Tarling, D.L. and Mitchell, J.G., 1976. Revised Cenozoic polarity time scale. Geology, 4: 133—136.
Taylor, D.J., 1971. Foraminifera and the Cretaceous and Tertiary depositional history in the Otway Basin in Victoria. In: H. Wopfner and J.G. Douglas (Editors), The Otway Basin of Southeastern Australia. Geol. Survs. South Aust. Vict., Spec. Bull., pp.217—233.
Taylor, L.W.H. 1975. Depositional and tectonic patterns in the western Coral Sea. In: D.A. Falvey and G.H. Packham (Editors), Southwest Pacific Workshop. Aust. Soc. Explor. Geophys., Bull., 6 (2/3): 33—35.

Trümpy, R., 1973. The timing of orogenic events in the central Alps. In: K.A. de Jong and R. Scholten (Editors), Gravity and Tectonics. Wiley, New York, N.Y., pp.229–251.
Van Andel, T.H., 1975. Mesozoic/Cenozoic calcite compensation depth and the global distribution of calcareous sediments. Earth Planet. Sci. Lett., 26: 187–194.
Van Andel, T.H., Heath, G.R. and Moore, T.L., 1975. Cenozoic tectonics, sedimentation and paleoceanography of the central equatorial Pacific. Geol. Soc. Am., Spec. Pap.
Veevers, J.J. and Johnstone, M.H., 1974. Comparative stratigraphy and structure of the Western Australian margin and the adjacent deep ocean floor. Initial Reports of the Deep Sea Drilling Project, 27. U.S. Govt. Printing Office, Washington, D.C., pp.571–585.
Veevers, J.J. and McElhinny, M.W., 1976. The separation of Australia from other continents. In: M.F. Glaessner (Editor), The Geosciences in Australia. Earth-Sci. Rev., 12(2/3): 139–159.
Vine, F.J., 1973. Continental fragmentation and ocean floor evolution during the past 200 m.y. In: D.H. Tarling and S.K. Runcorn (Editors), Implications of Continental Drift to the Earth Sciences, 2. Academic Press, London, pp.831–839.
Visser, W.A. and Hermes, J.J., 1962. Geological results of the exploration for oil in Netherlands New Guinea. Govt. Printing Office, The Hague, 265 pp.
Weissel, J.K. and Hayes, D.E., 1972. Magnetic anomalies in the southeast Indian Ocean. In: D.E. Hayes (Editor), Antarctic oceanology II: the Australian–New Zealand sector. Am. Geophys. Union, Ant. Res. Ser., 19: 165–196.
Weser, O.E., 1974. Sedimentological aspects of strata encountered on Leg 23 in the northern Arabian Sea. Initial Reports of the Deep Sea Drilling Project, 23. U.S. Govt. Printing Office, Washington, D.C., pp.503–519.
Whitmarsh, R.B., Hamilton, N. and Kidd, R.B., 1974. Paleomagnetic results for the Indian and Arabian Plates from the Arabian Sea cores. Initial Reports of the Deep Sea Drilling Project, 23. U.S. Govt. Printing Office, Washington, D.C., pp.521–525.
Willcox, J.B., 1976. The stratigraphy and structural evolution of the Great Australian Bight Basin. BMR Symp., 5th, Abstr., BMR J. Aust. Geol. Geophys., 1: 258–259.
Wopfner, H., 1974. Post-Eocene history and stratigraphy of northeastern South Australia. Trans. R. Soc. South Aust., 98: 1–12.
Wopfner, H., Callen, R. and Harris, W.K., 1974. The Lower Tertiary Eyre Formation of the south-western Great Artesian Basin. J. Geol. Soc. Aust., 21: 17–52.

Marine Geology, 26 (1978) 41—48

FURTHER EVIDENCE FOR THE EARLY CRETACEOUS BREAKUP OF GONDWANALAND OFF SOUTHWESTERN AUSTRALIA*

RUDI G. MARKL

Lamont-Doherty Geological Observatory of Columbia University, Palisades, N.Y. 10964 (U.S.A.)

(Received March 28, 1977)

ABSTRACT

Markl, R.G., 1978. Further evidence for the Early Cretaceous breakup of Gondwanaland off southwestern Australia. Mar. Geol., 26: 41—48.

Magnetic anomalies adjacent to the northwest margin of the Naturaliste Plateau are identified as Early Cretaceous reversals M-0 through M-4. These lineations indicate that northwest—southeast sea-floor spreading began off the present Naturaliste Plateau circa 118 m.y. B.P., 8 m.y. later than along the adjacent continental margin off Perth. The Naturaliste and Perth lineation sequences are offset 130 km dextrally by a fracture zone which passes near DSDP Site 257. The data indicate that an intracratonic small ocean basin existed seaward of Perth from about 126 to 118 m.y. B.P.

INTRODUCTION

Recent studies have tended to corroborate the basic DuToit (1937) reconstruction of eastern Gondwanaland, but with an important addition, namely, that pre-drift India extended eastward to Australia. The concept of such an enlarged, or "Greater" India implies that the early drift of India cannot have been exclusively northward, as previously assumed, but must have been northwestward, as suggested by Falvey (1972), Johnstone et al. (1973), Markl (1974a, b), Curray and Moore (1974), Veevers and Heirtzler (1974) and Johnson et al. (1976). Northwest—southeast sea-floor spreading off Western Australia was demonstrated by the magnetic anomaly investigations of Falvey (1972), Markl (1974b) and Larson (1975). These results, corroborated by basement ages determined from Deep Sea Drilling Project (DSDP) Legs 26 and 27 (Davies et al., 1974; Veevers et al., 1974) showed that continental dispersal began off northwestern Australia in Late Jurassic times, and off the southwest coast in Early Cretaceous times. The present study reports additional M-series (Larson and Pitman, 1972) anomalies indicative of northwest—southeast spreading off southwestern Australia.

---

*L-DGO Contribution No. 2565.

A two-week cruise of R/V "Vema" (3304) was devoted to surveying the Naturaliste fracture zone (Markl, 1974a, b, c) and attempting to determine from morphologic features and magnetic anomalies whether lateral offset exists across the fracture zone. A premise of this search was that the region bounded by the Naturaliste and Diamantina fracture zones was, like the region north of the plateau, formerly occupied by some part of Greater India and, therefore, might yield Mesozoic-series anomalies corresponding to those of the Perth sequence (Markl, 1974b).

MAGNETIC ANOMALY DATA

A long magnetic profile obtained parallel to and *west* of the Naturaliste fracture zone revealed no identifiable anomalies, nor any correlations to the few suitably oriented existing magnetic profiles. However, several tracklines parallel to and *east* of the Naturaliste fracture zone did reveal Mesozoic anomalies M-0 through M-4, adjacent to the northwest margin of the Naturaliste Plateau (Fig.1). The character of the M-4 anomaly, shown in Fig.2, profile 3304-1, is perturbed by a topographic spur; likewise, anomaly M-4 on profile 3304-3 may be influenced, although less so, by minor relief of the plateau margin. However, there is no disturbing relief on profile 3304-2, which displays the best-developed M-4 anomaly. Existing tracklines R/V "Robert D. Conrad" 1106 and USNS "Eltanin" 48 help define the trend of anomaly M-0 (although they intersect right over it), but anomaly M-0 cannot be positively identified on D/V "Glomar Challenger" cruise 26, nor are anomalies M-2 and M-4 of this sequence observed on GC26. The (half) spreading rate between anomalies M-0 and M-4 is 1.8 cm/yr, which is identical to the average rate of the Perth sequence when the revised time scale of Larson and Hilde (1975) is used.

It is apparent from the dextral offset of approximately 130 km between corresponding lineations of the "Naturaliste" anomaly sequence and the Perth sequence, that a northwest—southeast-striking fracture zone exists between these sequences. Unlike the Naturaliste fracture zone, which is in all respects a prominent feature, the magnetic, gravity, and seismic reflection data provide no obvious indication of the precise location of the fracture zone (or zones). The magnetic anomaly data near the plateau would permit the fracture zone to lie anywhere within a band extending about 35 km to either side of the GC26 track. DSDP Site 257, in the Cretaceous quiet zone, appears to lie very near the inferred projection of the fracture zone; the age of the basal sediments at Site 257 is approximately 102 m.y. (Davies et al., 1974). The spreading rate between Site 257 and anomaly M-2 of the Naturaliste sequence is 1.6 cm/yr, whereas the rate computed between Site 257 and anomaly M-2 of the Perth sequence is 2.6 cm/yr (also using the Larson and Hilde (1975) time scale). Both 1.6 and 2.6 are plausible rates, considering present knowledge of global changes in spreading rates in the mid-Cretaceous, so this exercise does not specify whether Site 257 lies east or west of the fracture zone. Acoustic basement dips southeast in a broad

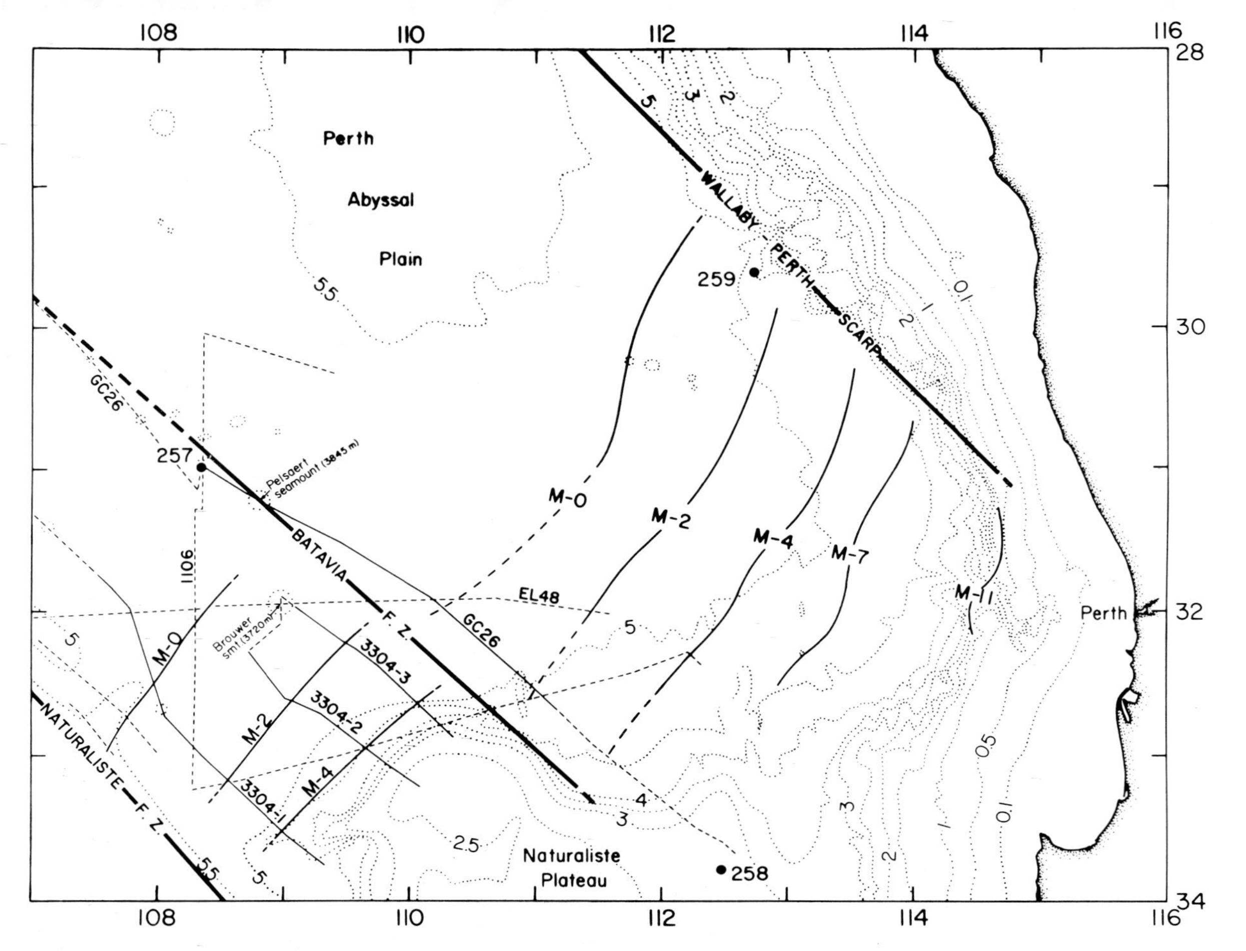

Fig.1. The Naturaliste and Perth Early Cretaceous magnetic lineation sequences (separated by the Batavia fracture zone). Ship's track segments shown by solid lines correspond to the magnetic profiles in Fig.2; other tracklines in the area are dashed. DSDP sites are indicated by large dots. Isobaths (dotted lines) are labelled in kilometers.

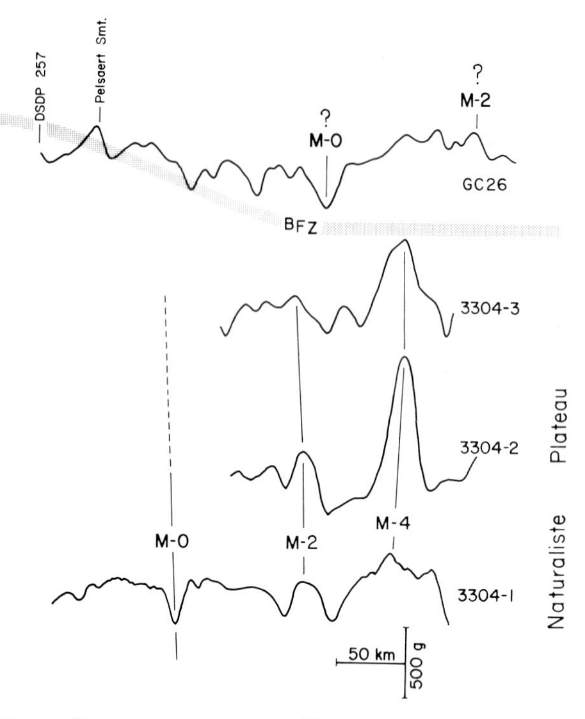

Fig.2. Observed magnetic profiles of the Naturaliste sequence ("Vema" 3304) projected perpendicular to the strike of the lineations. Trackline GC26 crosses the Batavia fracture zone (*BFZ*) obliquely and does not clearly show M-series anomalies of either the Perth or the Naturaliste sequence.

region surrounding Site 257 (Markl 1974a; 1978). The basement contours indicate a 125-km dextral offset just west of the GC26 track which matches the offset of the magnetic lineations. Also, a steep northwest—southeast-trending scarp exists where the fracture zone is inferred to intersect the plateau. The few basement highs and seamounts known in the area show no pattern which might define the trace of the fracture zone, although Pelsaert seamount lies on its probable trace (Fig.1). This morphologic evidence, in conjunction with the constraints of the magnetic anomaly data, the requirement for parallelism with the Naturaliste fracture zone, and the assumption that relatively constant spreading rates obtained in the adjacent spreading compartments suggests that this fracture zone, henceforth referred to as the "Batavia" fracture zone*, passes east of Site 257.

*Dutch merchant ships bound for Batavia (Djakarta) are the first known to have sailed these waters. They followed the trade winds eastward from Africa, a route pioneered by Henrik Brouwer in 1611, turning northward west of Australia. The difficulty of determining longitude caused many ships, including one named "Batavia", to be wrecked on the Australian coast. The Batavia's captain, Francois Pelsaert, sailed 1800 miles to Java in a small boat, returning to find that mutineers had murdered 125 of the survivors (Joy, 1971; Green, 1975).

INTERPRETATION OF DATA

The nature of the crust of the Naturaliste Plateau has yet to be established. Some evidence suggests it is continental (Francis and Raitt, 1967; Heezen and Tharp, 1973; Petkovic, 1975), and other evidence has been interpreted to imply an oceanic origin (Luyendyk and Davies, 1974). An additional unknown factor is whether the plateau has always been fixed in its present relationship to Australia. Markl (1974a) speculated that the plateau may have moved northwest, initially accompanying the Indian plate; Veevers et al. (1975) invoked such a translation.

The magnetic anomaly data show that sea-floor spreading began off the present northwestern margin of the Naturaliste plateau in earliest Barremian times (ca. 118 m.y. B.P.), approximately 8 m.y. later than off the nearby Australian margin, where spreading began in late Valanginian times. It seems highly improbable that accretion was not occurring at some location within the Naturaliste spreading compartment (that bounded by the Naturaliste and Batavia fracture zones and their projections) in this 8-m.y. period, during which approximately 290 km of oceanic crust were generated in the adjacent Perth compartment (at a total rate of 3.6 cm/yr). Although opening between India and Western Australia began in the north and progressed southward (Veevers et al., 1971; Markl, 1974b; Larson, 1975), the apparent disparity in onset of spreading in the Naturaliste Plateau area cannot be explained by this means.

Conceptually, there appear to be three alternative ways to account for the disparity: (1) spreading within the Naturaliste compartment occurred somewhere to the northwest, between the plateau and the northern edge of Greater India, (2) spreading occurred to the southeast, between the plateau and Antarctica, and (3) accretion occurred within the area presently occupied by the plateau. The first two alternatives imply subsequent ridge jumps to the northwest margin of the plateau just prior to M-4 time (118 m.y. B.P.).

Alternative (1) presupposes that the Naturaliste Plateau is continental and was fixed to Australia as now; it probably precludes the existence of a rift between the plateau and Antarctica in Early Cretaceous times. According to Veevers et al. (1975), the Tethyan margin of Greater India lay approximately 1800 km northwest of the Naturaliste Plateau. Although magnetic anomaly data (Markl, 1974b; Larson, 1977) document an offset of at least 1000 km along the Wallaby—Perth scarp (transform fault), it is geologically unlikely that the 400-km-wide Perth compartment was bounded on the west by an even longer fault. It is probable that the pre-anomaly M-4 phase of spreading occurred nearer the plateau, or even at its northwestern edge. The bathymetry of the southern Wharton Basin (Markl, 1974c) shows a disproportionate number of seamounts and larger elevations between the projected traces of the Naturaliste and Batavia fracture zones. These include the plateau-like northern extremity of Broken Ridge and a long ridge near 30°S 105°E that subsequent surveying (Markl, 1978) suggests is a volcanic pile. The few magnetic profiles adjacent to this ridge show no identifiable anomalies, but it is conceivable that it originated at the northwestern edge of the Naturaliste Plateau.

Alternative (2), that initial spreading in the Naturaliste compartment occurred south of the Naturaliste Plateau also presumes a continental origin for the plateau, but implies that it formerly lay about 290 km to the southeast, having been carried to its present location as part of the Indian plate. Marked northwest—southeast morphologic trends on the plateau and adjacent Australian margin, together with the observation that the Naturaliste fracture zone extends approximately 200 km southeast of the plateau first raised this possibility (Markl, 1974a). Magnetic profiles between the Naturaliste Plateau and Diamantina fracture zone are dominated by a pair of (unidentified) positive anomalies of larger amplitude (500 gamma) than occur within or south of the Diamantina fracture zone. The anomalies can be correlated between 112° and 114°E, however, the 085° trend of the lineations suggests that they are unrelated to the Early Cretaceous spreading episode.

Alternative (3), the assumption that the Naturaliste Plateau is of oceanic origin, opens a range of possibilities depending upon when and how the plateau formed and whether it has been fixed to Australia. Cores from DSDP Sites 258 and 264 show it to have been a marine feature since at least 105 m.y. B.P. and above the lysocline since Late Cretaceous times (Davies et al., 1974; Hayes et al., 1975). Although the former authors discuss the possible Early Cretaceous origin of the plateau as the result of volcanic activity along the incipient rift between Australia and Antarctica, this is difficult to reconcile with the fact that normal oceanic crust was accreting simultaneously in the adjacent Perth compartment. One could postulate that the plateau represents normal oceanic crust (of M-11 to pre-M-4 age) that was subsequently uplifted. An origin of this type appears plausible for Broken Ridge, whose northern flank slopes gently to abyssal depths, but not for the Naturaliste Plateau, which rises abruptly from the deep basin where anomalies M-0 through M-4 occur.

Although the evidence is not conclusive, it is probable that the initial rift in the Naturaliste crustal compartment lay northwest of the plateau (alternative 1). However, all of the alternatives appear to have two important results in common. First, that the receding margin of India (in the Perth compartment) was abreast of the (present) northwest margin of the plateau at the time spreading began there. Second, that the crust generated in the Perth compartment prior to M-4 time formed an intracratonic small ocean basin (Fig.3).

Wherever the initial rift transected the Naturaliste compartment, it seems certain that it was transformed south-eastward again, along the trace of the Naturaliste fracture zone. The fact that the fracture zone extends considerably south of the plateau and northeast—southwest basement trends are observed in the triangular area between it and the Diamantina fracture zone suggests that some part of India, rather than Antarctica, formerly lay southwest of the Naturaliste Plateau.

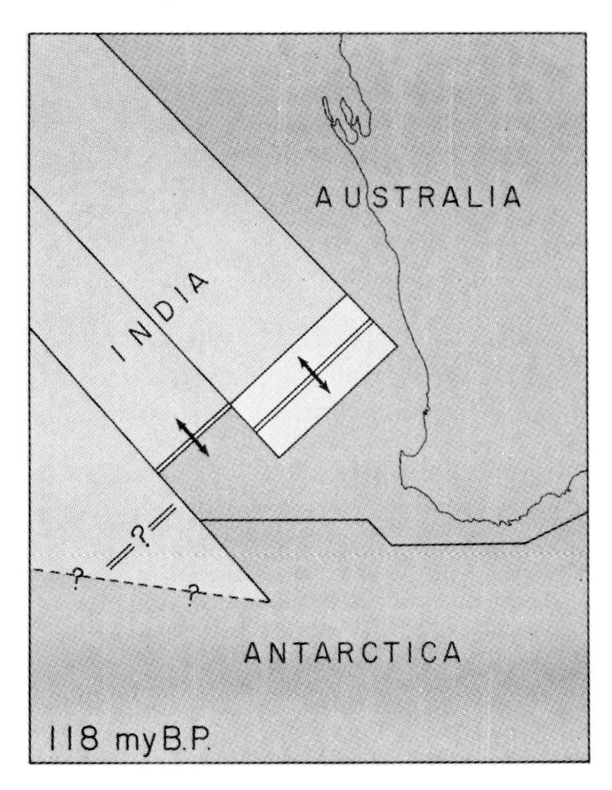

Fig.3. Diagrammatic sketch of the junction of Greater India and Australia—Antarctica at the time spreading began off the northwest margin of the Naturaliste Plateau. An intracratonic small ocean basin formed in the Perth compartment during the preceding 8 m.y.

ACKNOWLEDGEMENTS

I thank the complement of "Vema" cruise 3304 for their cooperative spirit and professionalism, and S. Cande for review. Data collection and reduction was supported by National Science Foundation grant OCE 76-01434 to Lamont-Doherty Geological Observatory of Columbia University.

REFERENCES

Curray, J.R. and Moore, D.G., 1974. Sedimentary and tectonic processes in the Bengal deep-sea fan and geosyncline. In: C.A. Burk and C.L. Drake (Editors), The Geology of Continental Margins. Springer, New York, N.Y., pp.617—627.

Davies, T.A., Luyendyk, B.P. et al., 1974. Initial Reports of the Deep Sea Drilling Project, 26. U.S. Govt. Printing Office, Washington, D.C.

DuToit, A.L., 1937. Our Wandering Continents — An Hypothesis of Continental Drifting. Oliver and Boyd, London, 366 pp.

Falvey, D.A., 1972. Sea-floor spreading in the Wharton basin (northeast Indian Ocean) and the breakup of eastern Gondwanaland. J. Aust. Pet. Explor. Assoc., 12(2): 86—88.

Francis, T.J.G. and Raitt, R.W., 1967. Seismic refraction measurements in the southern Indian Ocean. J. Geophys. Res., 72: 3015—3041.
Green, J.N., 1975. The VOC ship "Batavia" wrecked in 1629. Int. J. Naut. Archeol. Underwater Explor., 4(1): 43—63.
Hayes, D.E., Frakes, L.A. et al., 1975. Initial Reports of the Deep Sea Drilling Project, 28. U.S. Govt. Printing Office, Washington, D.C., pp.19—27.
Heezen, B.C. and Tharp, M., 1973. USNS "Eltanin" cruise 55. Antarct. J. U.S., 8(3): 137—141.
Johnson, B.D., Powell, C. McA. and Veevers, J.J., 1976. Spreading history of the eastern Indian Ocean, and Greater India's northward flight from Antarctica and Australia. Geol. Soc. Am., Bull., 87(11): 1560—1566.
Johnstone, M.H., Lowry, D.C. and Quilty, P.G., 1973. The geology of southwestern Australia — a review. J.R. Soc. West. Aust., 56: 5—15.
Joy, W., 1971. The Explorers. Rigby, Australia, 159 pp.
Larson, R.L., 1975. Late Jurassic sea-floor spreading in the eastern Indian Ocean. Geology, 3: 69—71.
Larson, R.L., 1977. Early Cretaceous breakup of Gondwanaland off Western Australia. Geology, 5: 57—60.
Larson, R.L. and Hilde, T.W.C., 1975. A revised time scale of magnetic reversals for the Early Cretaceous and Late Jurassic. J. Geophys. Res., 80(17): 2586—2594.
Larson, R.L. and Pitman III, W.C., 1972. World-wide correlation of Mesozoic magnetic anomalies, and its implications. Geol. Soc. Am., Bull., 83: 3645—3662.
Luyendyk, B.P. and Davies, T.A., 1974. Results of DSDP leg 26 and the geologic history of the southern Indian Ocean. Initial Reports of the Deep Sea Drilling Project, 26. U.S. Govt. Printing Office, Washington, D.C., pp.909—943.
Markl, R.G., 1974a. Bathymetry, Sediment Distribution, and Sea-Floor Spreading History of the Southern Wharton Basin, Eastern Indian Ocean. Thesis, Univ. of Connecticut, Storrs, Conn. (unpublished).
Markl, R.G., 1974b. Evidence for the breakup of eastern Gondwanaland by the Early Cretaceous. Nature, 251 (5472): 196—200.
Markl, R.G., 1974c. Bathymetric map of the eastern Indian Ocean (southern Wharton basin). In: T.A. Davies, B.P. Luyendyk et al., Initial Reports of the Deep Sea Drilling Project, 26. U.S. Govt. Printing Office, Washington, D.C., pp.967—968.
Markl, R.G., 1978. Basement morphology and rift geometry near the former junction of India, Australia, and Antarctica, in press.
Petkovic, P., 1975. Origin of the Naturaliste plateau. Nature, 253: 30—33.
Veevers, J.J., Jones, J.G. and Talent, J.A., 1971. Indo-Australian stratigraphy and the configuration and dispersal of Gondwanaland. Nature, 229: 383—388.
Veevers, J.J. and Heirtzler, J.R., 1974. Tectonic and paleogeographic synthesis of leg 27. In: J.J. Veevers, J.R. Heirtzler et al., Initial Reports of the Deep Sea Drilling Project, 27. U.S. Govt. Printing Office, Washington, D.C., pp.1049—1054.
Veevers, J.J., Heirtzler, J.R. et al., 1974. Initial Reports of the Deep Sea Drilling Project, 27. U.S. Govt. Printing Office, Washington, D.C.
Veevers, J.J., Powell, C.McA. and Johnson, B.D., 1975. Greater India's place in Gondwanaland and in Asia. Earth Planet. Sci. Lett., 27: 383—387.

Marine Geology, 26 (1978) 49—70

INDIAN OCEAN SEDIMENT DISTRIBUTION SINCE THE LATE JURASSIC

ROBERT B. KIDD and THOMAS A. DAVIES

Institute of Oceanographic Sciences, Wormley, Surrey (Great Britain)

Department of Geology, Middlebury College, Middlebury, Ve. (U.S.A.)

(Received March 28, 1977)

ABSTRACT

Kidd, R.B. and Davies, T.A., 1978. Indian Ocean sediment distribution since the Late Jurassic. Mar. Geol., 26: 49—70.

Cores obtained by deep sea drilling in the Indian Ocean provide a sedimentary record from which are deduced changing patterns of sedimentation through the Late Mesozoic and Cenozoic. Comparisons between: (1) empirical subsidence curves and sediment sequences at individual sites; and (2) paleobathymetric reconstruction maps and past sediment distributions, convincingly demonstrate the interrelationship of sedimentary and tectonic development within this the most recently formed of the major oceans.

INTRODUCTION

The broad picture of changing paleogeography in the Southern Hemisphere was documented by McKenzie and Sclater in 1971, when they used the distribution of oceanic magnetic lineations and the paleomagnetism of the surrounding continents to reconstruct the evolution of the Indian Ocean since the Late Cretaceous. The results of Legs 22 to 27 of the Deep Sea Drilling Project, which drilled more than fifty Indian Ocean holes in 1972 (Fig.1), allowed considerable refinement of these earlier paleogeographic reconstructions. More importantly perhaps, the drilling also provided our best record to date of changing distributions of deep sea sediments which came about during the evolution of a major ocean basin. The purpose of this contribution is to briefly document changing patterns of Indian Ocean sedimentation through the Late Mesozoic and Cenozoic and to compare these with reconstructions of paleobathymetry and paleogeography. (Fuller documentation of the distributions through time, based on deep sea drilling results, appears in Davies and Kidd, 1977.) We then highlight some significant features which yield information on paleoenvironmental conditions both within the ocean and in its surrounding land regions.

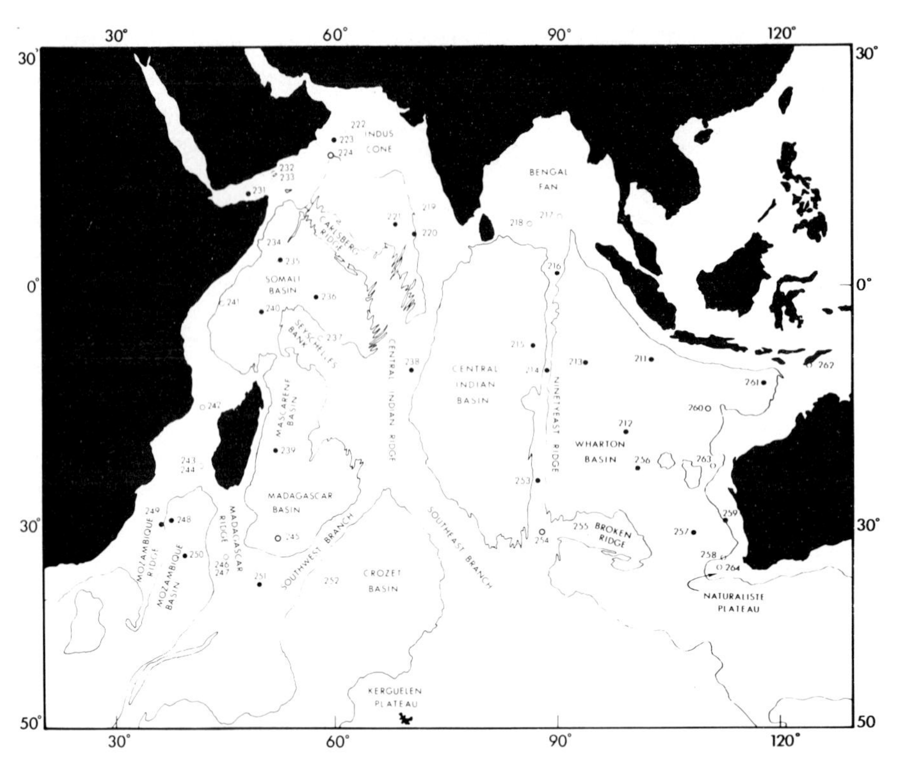

Fig.1. Deep Sea Drilling Project sites in the Indian Ocean (solid dots = sites which penetrated to oceanic basement; open circles = other sites; contour at 4000 m water depth).

Present-day sedimentation

Distributions of bottom sediments within the Indian Ocean, based on all coring data collected between 1872 and 1970, were mapped for the International Indian Ocean Expedition atlas (Udintsev, 1975, pp.125–135). Carbonate sediments are accumulating in the shallow areas along the African and Australian coasts, on the shallow ridges and platforms of the western Indian Ocean, on Ninetyeast and Broken Ridges and in the shallower parts of the Somali Basin (Fig.2).

In the deep oceans, the carbonate compensation depth is defined as the level at which carbonate input from the surface is balanced by dissolution. Its trace on the sea floor represents a boundary or "carbonate line", an analogy to the "snow line" on land, which separates calcareous oozes from clays. On the basis of piston coring, the carbonate compensation depth, or CCD, lies deeper than 4000 m in the Indian Ocean. It probably averages 4500 m water depth as in the other oceans (Berger and Winterer, 1974), with shoaling towards the continents and deepening to around 5000 m below the equatorial high-productivity zone (Kolla et al., 1976).

Below the CCD, the deep basins are receiving deep sea clays and siliceous sediments, with the latter being confined to the equatorial and sub-polar

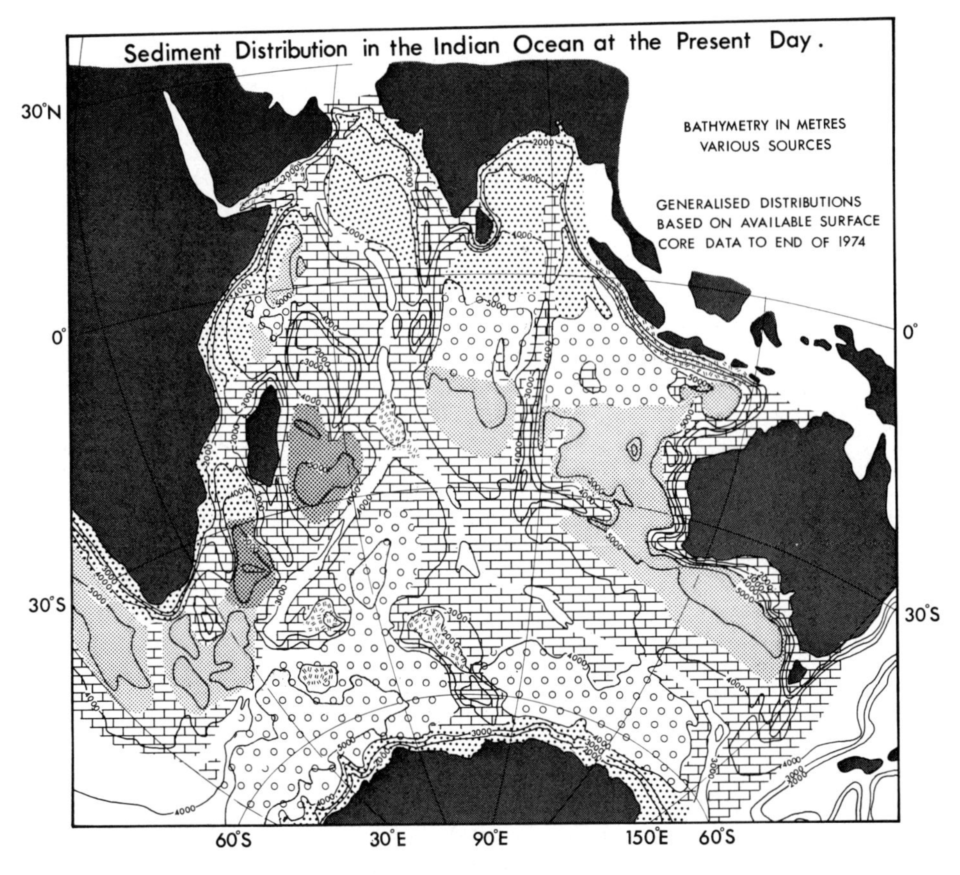

Fig.2. Sediment distribution in the Indian Ocean at the present day (based on available coring data to the end of 1974). Key to symbols: bricks = calcareous ooze; open circles = siliceous ooze; dots = terrigenous sediments; light stipple = pelagic clay; dark stipple = other types of deep sea clay; double dashes = volcanogenic sediments.

high-productivity regions. In the central Wharton Basin, the southern Mascarene Basin, and parts of the Crozet Basin, sedimentation appears to have stopped and deposition is minimal between Broken Ridge and the southeast Indian Ridge, where an extensive area of manganese nodule pavement is present (Kennett and Watkins, 1975). The lack of deposition in these areas is presumed to be due to isolation and/or the activity of deep geostrophic currents.

Superimposed on this relatively simple pattern of oceanic sedimentation is the influence of land-derived sediments. Thick terrigenous sediment accumulations are associated with areas where some of the world's major rivers enter the ocean. Thus, the huge Indus and Bengal submarine fans result from erosion of the youthful mountain ranges of the Indian sub-continent and the Zambesi Fan receives sediments from a wide area of southern Africa. The major present-day development of volcanogenic sediment lies south of the Indonesian island arc. The arc contains 14% of the world's active volcanoes whose products are silicic in composition, and whose activity is explosive in nature, thus causing widespread dispersal of pyroclastic deposits.

DATA: PROCESSING AND PRESENTATION

Sources of data

This account is based largely on data from those volumes of the Initial Reports of the Deep Sea Drilling Project that refer to the Indian Ocean drilling (Von der Borch et al., 1974; Whitmarsh et al., 1974; Fisher et al., 1974; Simpson et al., 1974; Davies et al., 1974; Veevers et al., 1974; Hayes et al., 1975). The reader is referred to these original sources for detailed core descriptions. During the course of this study, we were fortunate that both of us were based at the DSDP West Coast repository. Thus we were able to re-examine large numbers of the stored cores and to resolve any problems encountered, such as those caused by Leg to Leg differences in sediment terminology. Therefore we stress that some detailed aspects, most often those relating to sediment classification, may not necessarily agree with the published descriptions.

Lithologic data

Our approach has been to map generalised paleodistributions by reducing the large number of sediment types described from the drillsites into five broad lithologic classes based on composition: terrigenous (clastic, detrital) sediments (including shallow-water bioclastic limestones), volcanogenic sediments, calcareous ooze, siliceous ooze, and clays. The clays are further subdivided into "pelagic" clays and "other types" of deep sea clays. This distinction is made primarily on accumulation rate, but colour, composition, organic carbon content and bedding features are also taken into account. We consider that "pelagic" deep sea clays accumulated at rates less than 0.5 cm per thousand years, have no bedding features, or significant coarse fraction, and have compositions indicative of extremely slow accumulation, including negligible amounts of organic carbon and pyrite. Distinguishing these pelagic clays from the other types, which do not fit these criteria, helps draw attention to some features of past Indian Ocean sedimentation that might otherwise be masked. Volcanogenic clay sediments do not appear with the "other types" of deep sea clay. This is because the separate volcanogenic category includes both volcanic ashes and other sediments which, according to criteria based on the relative abundances of diagnostic clay minerals, can be considered volcanogenic in origin (Vallier and Kidd, 1977).

Paleontologic data

In assigning ages to the sediments, we have accepted those stated in the Initial Report volumes of DSDP. Where radiometric ages of igneous rocks were available, they have been used in assigning ages to the bases of drilled sections, in preference to the paleontologic age of sediments immediately overlying oceanic basement. Absolute chronology for the Cenozoic sediments is from the time scales of Berggren (1972), or more recent revisions by Berggren and Van Couvering (1974) or Ryan et al. (1974); chronology for the Mesozoic is as used on Leg 27 (Veevers et al., 1974). The frequency of continuous or near-continuous coring during the Indian Ocean program means that large unsampled intervals between cores are relatively infrequent and consequently that our knowledge of the time distribution of the Indian Ocean sediments is more complete than in the other oceans. Major gaps in the record can here be more confidently assigned to sedimentary hiatuses (Davies et al., 1974).

Presentation techniques

A number of techniques are possible for the presentation of the lithologic data against time (Kidd and Weser, 1974, 1975; Davies et al., 1975). Two which display the lithologies against a background of changing paleobathymetry are employed here.

Sclater et al. (1971) showed the empirical relationship that exists between depth and age of the oceanic crust, caused by thermal contraction of the cooling lithosphere as sea-floor spreading moves it away from the spreading centre. A sinking curve for the Indian Ocean, compiled from the basement ages of DSDP drillholes together with independent geophysical data, was constructed by Luyendyk and Davies (1974) (Fig.3). Such a curve may be used directly in reconstructing individual subsidence curves for each of the Indian Ocean sites that reached igneous "basement". (We exclude sites which were terminated in igneous rock thought to represent sills.) For these sites the

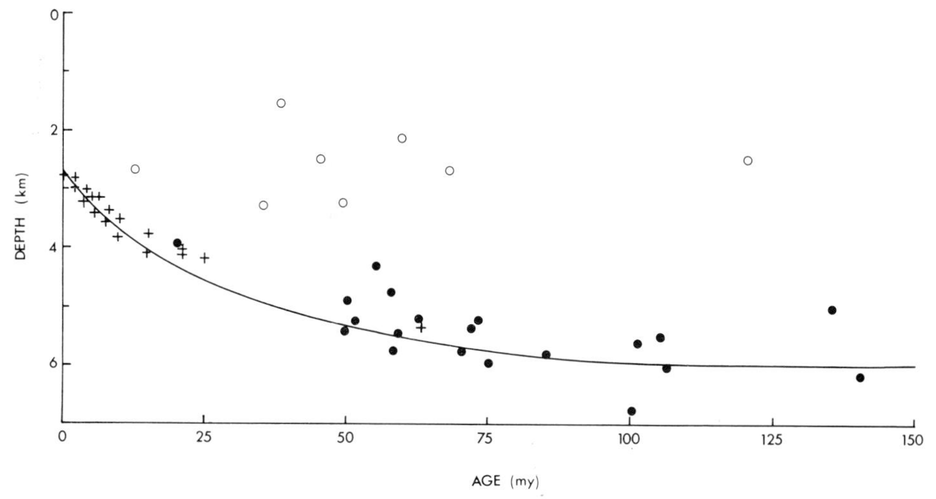

Fig.3. Age/depth curve for the Indian Ocean. Crosses = geophysical data from Sclater et al., 1971; solid dots = Deep Sea Drilling Project basement ages from "normal" ocean crust; open circles = Deep Sea Drilling Project "basement" ages from obviously anomalous regions such as aseismic ridges (from Luyendyk and Davies, 1974).

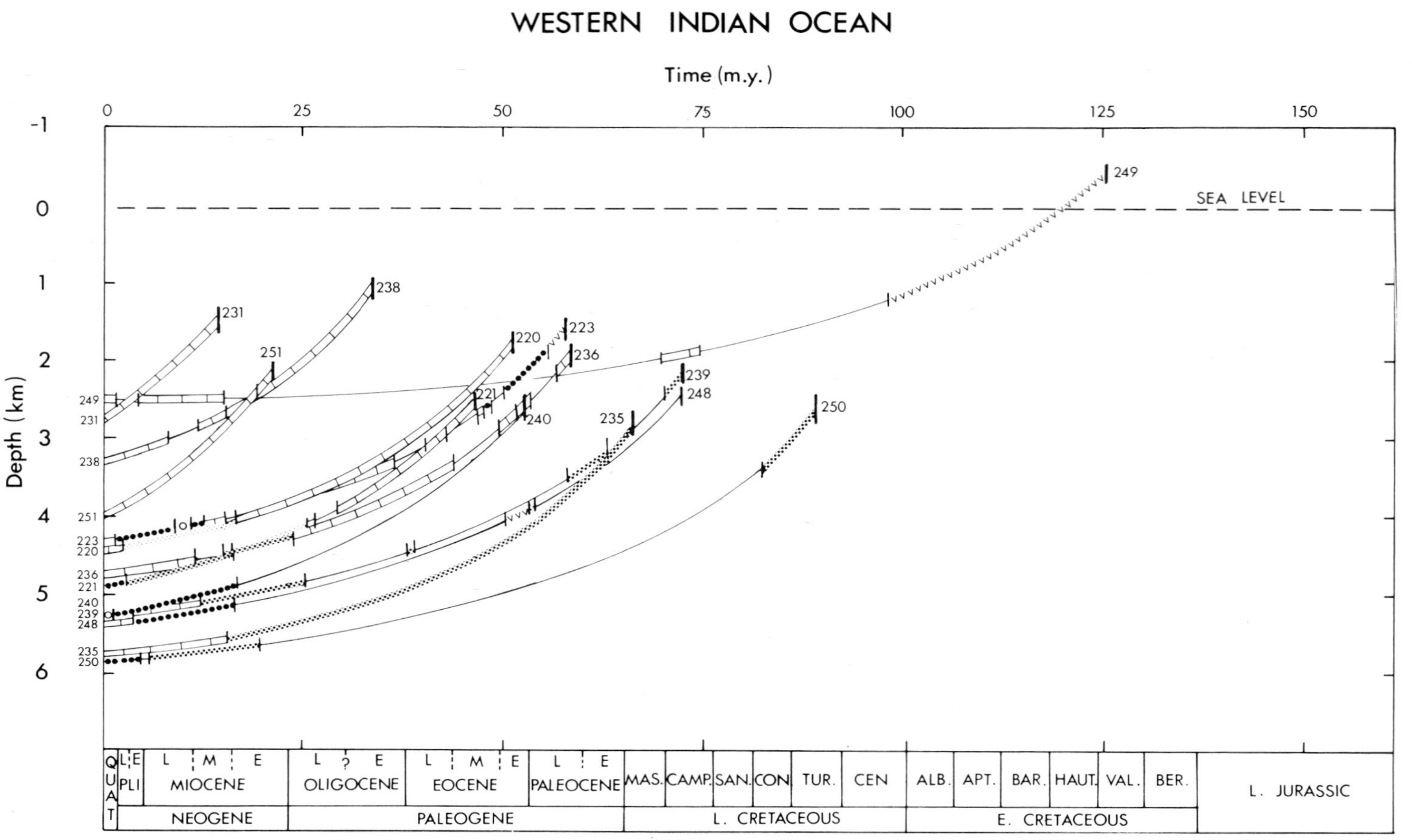

Fig.4. Paleodepth curves and lithologies for sites in the Western Indian Ocean; vvvv = volcanogenic sediments; bricks = calcareous ooze; light stipple = pelagic clays; dark stipple = other types of deep sea clay; solid dots = terrigenous sediments; open circles = siliceous ooze; thin curving line = hiatus.

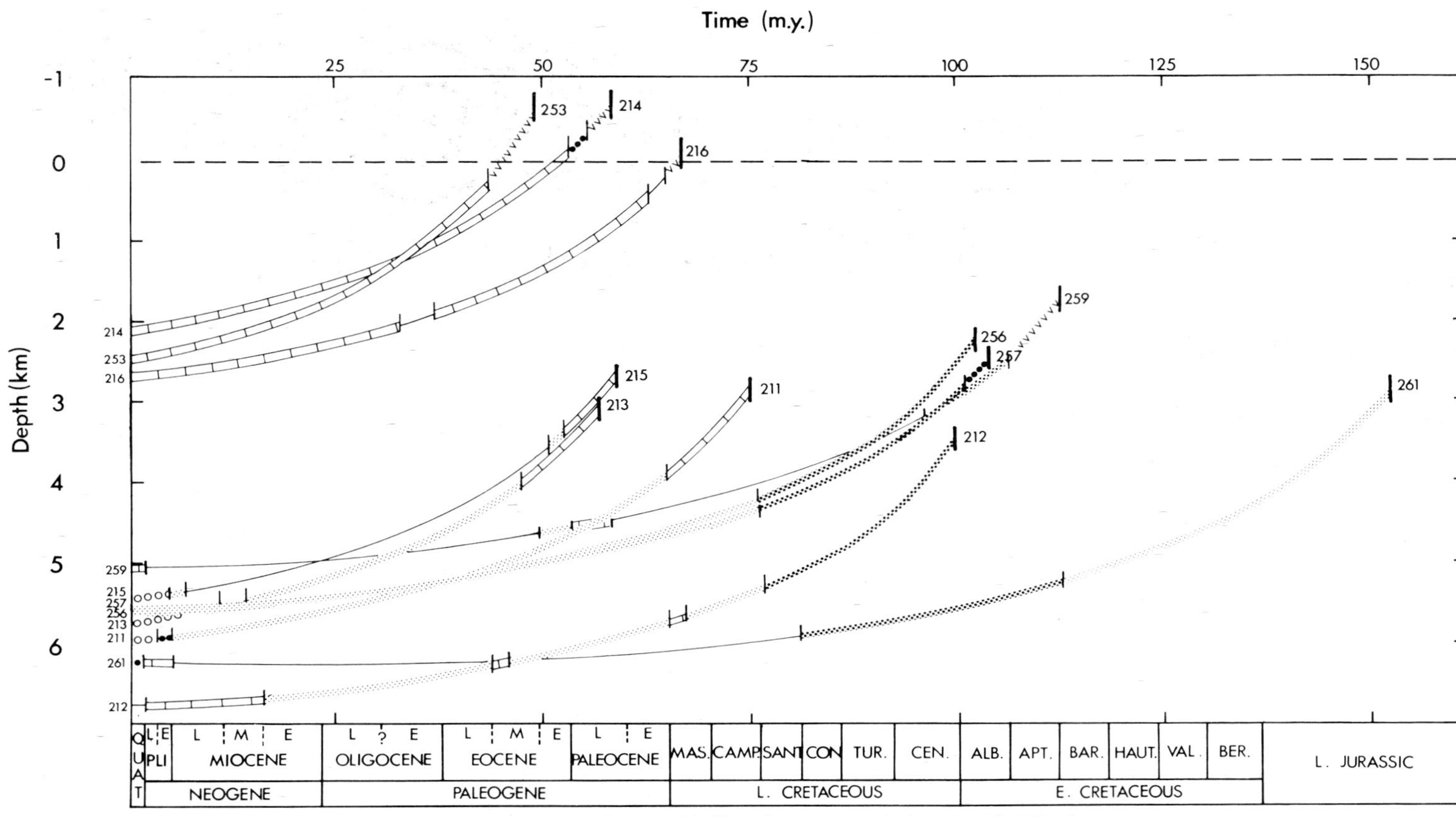

Fig.5. Paleodepth curves and lithologies for sites in the Eastern Indian Ocean; symbols are as in Fig.4.

present-day depth of the basement is known (water depth plus sediment thickness). Thus the ocean crust at each site may be traced (or "backtracked") parallel to the average subsidence curve for the whole ocean (Fig.3), now with a reversed time scale. This parallel trace is terminated at the age of formation deduced for the piece of oceanic crust. Figs.4 and 5 show individual subsidence curves for the Indian Ocean sites on which are plotted the lithologic successions encountered in the drillholes. Also recorded are the hiatuses present. For example, in Fig.5 the basement of Site 215, presently at a depth of 5492 m (5321 m water depth plus 151 m sediment thickness) "backtracks" through a succession of progressively older sediments (siliceous ooze—pelagic clay with hiatus—calcareous ooze) to a late Paleocene age of formation at a paleodepth of around 2800 m.

A second method of presenting the lithologic data is to plot the spatial distributions of sediment type on reconstructions of past Indian Ocean geography, based on magnetic data, and past bathymetry, as predicted from the subsidence relationship.

Sclater et al. (1977) have constructed preliminary paleobathymetric charts of the Indian Ocean for 36 m.y.B.P., 53 m.y.B.P. and 70 m.y.B.P. They employ for these reconstructions.

(1) The depth versus age relationship for typical ocean crust in conjunction with magnetic anomaly distributions; they show that, after its creation at a spreading centre at a mean depth of 2500 ± 300 m, ocean crust subsides to around 3000 m after about 2 m.y., 4000 m after 20 m.y., and 5000 m after 50 m.y. Thus the 4000-m depth contour at 36 m.y.B.P. can be traced from the present-day position of magnetic anomaly 24 which represents oceanic crust 56 m.y. old; the 5000-m contour is traced from the 86-m.y. isochron and so on.

(2) A basic assumption that all the aseismic ridges were created at sources some altitude above that normal for the spreading ridge, but subsided away at the same rate as adjacent oceanic crust: this is the case for Ninetyeast Ridge, the best-known aseismic ridge, which formed at sea level, (Pimm, 1974; Luyendyk and Davies, 1974), and also for other aseismic ridges for which sedimentary successions are established (Thiede et al., 1975).

(3) Some improvements and extensions of the tectonic history as proposed in the paleogeographic reconstructions by McKenzie and Sclater in 1971, these based on more recent magnetic anomaly and bathymetric data.

Since magnetic and tectonic data is sketchy for the western Indian Ocean prior to 50 m.y.B.P., Sclater et al. (1971) consider only their Early Oligocene (36 m.y.B.P.) reconstruction to be quantitative, while those for the Early Eocene (53 m.y.B.P.) and Late Cretaceous, Maastrichtian to Campanian (70 m.y.B.P.) may be "highly speculative, especially between Africa and Antarctica".

In Figs.6, 7 and 8, we plot the distributions of our five broad sediment classes as found in the DSDP holes, on the paleobathymetric reconstructions. This method allows us to test the validity of the reconstructions (particularly the early ones) from the distribution of carbonate and non-carbonate sediments or of proven shallow-water deposits. On the other hand, the reconstructions provide a framework for extrapolation of the mapped distributions between drillsites.

COMPARISON OF SUBSIDENCE CURVES AND SEDIMENT SEQUENCES

A common sequence of lithologies deposited through the subsidence history of DSDP drillsites is for calcareous ooze, deposited on the young oceanic crust, to be replaced by pelagic clay, if the site sank below the CCD (Berger and Winterer, 1974, p.39, fig.21). None of the curves for the Indian Ocean sites show progressions as straightforward as this, but in most cases explanations are relatively simple. Some sites, because of their youth or their "initiation" on an aseismic ridge, never reach the CCD level: Sites 231, 238 and 251 on Fig.4; Sites 253, 214 and 216 on Fig.5. At others, where the initiation of subsidence was accompanied by explosive volcanic activity, the first sediments are volcanogenic in character: Site 249 on Fig.4; Sites 253, 214, 216 and 259 on Fig.5. Once a site has sunk below the CCD as indicated by clay deposition, continuing sea-floor spreading might bring it below a zone of high siliceous plankton productivity so accumulating siliceous oozes: Sites 215 and 213 on Fig.5. Similarly a site coming under the influence of a major source of land-derived material, such as a submarine fan or continental margin, would accumulate terrigenous sediments: Site 221 on Fig.4, and even carbonates, where deposition rates were sufficiently rapid. Large-scale sediment slumping from plateaus or other topographic highs is a further process which can introduce carbonate sediments into clay regimes below the CCD: Site 212 on Fig.5 (Von der Borch et al., 1974).

Some of the Indian Ocean sites clearly began accumulating volcanogenic and/or terrigenous deposits at, or above, sea level. These are on the aseismic Ninetyeast (Sites 214, 216 and 253 on Fig.5) and Mozambique Ridges (Site 249 on Fig.4). Other sites began their sedimentary history deeper than the CCD of the time, and in many cases have remained below it throughout: Site 250 on Fig.4; Sites 256 and 257 on Fig.5. A distinction may be made between the levels at which the sites initiated in the Tertiary first accumulated clay sediments, and the levels at which this occurs for sites of Mesozoic origin. From this it would appear that the CCD during parts of the Mesozoic may have ranged up to 2000 m, that is, 2500 m above its present-day average depth. The curves also illustrate well the considerable intervals of time represented in the sediment columns by sedimentary hiatuses. Many, which clearly were synchronous, were developed at a wide range of water depths.

COMPARISON OF PALEOBATHYMETRIC RECONSTRUCTIONS AND EVOLVING SEDIMENT DISTRIBUTIONS

Mesozoic sediments

Mesozoic sediments were encountered at relatively few localities, principally around the ocean margins. The Jurassic is represented in only one hole

(Site 261) northwest of Australia, where the sediments are predominantly clays.

The Early Cretaceous is little better represented. Volcanogenic sediments were deposited on the Mozambique Ridge, while off Australia detrital clays were accumulating. These "other clays" contain significant amounts of pyrite and organic matter which suggest deposition under anoxic conditions. Further north and west at Site 260, siliceous sediments were sampled and radiolarites of similar age are known from the Carnarvon Basin in northwest Australia (Brown et al., 1968). On Naturaliste Plateau, Albian/Aptian detrital clays derived from the weathering of basaltic igneous rocks may be related to the mid-Cretaceous volcanic episode described by Brown et al. (1968) in the Perth Basin.

The separation of Africa from Antarctica commenced during, or prior to, the Early Cretaceous (Luyendyk, 1974) and was followed in Albian—Aptian times by the separation of India from Australia/Antarctica (Sclater and Fisher, 1974). These events may be recorded by the volcanogenic sediments on Mozambique Ridge and Naturaliste Plateau respectively. The separation of India from Australia is thought to have produced a narrow restricted basin, or series of basins, and we can show that the first infill was fine-grained detrital material. To the north lay open ocean and the sediments indicate that there siliceous plankton productivity was relatively high.

Late Cretaceous (Fig.6). More evidence is available on the distribution of Late Cretaceous sediment types. The early patterns of sedimentation in relatively small developing basins appear to have persisted throughout the Cretaceous and thus we can comment on the earliest paleobathymetric reconstruction at 70 m.y.B.P.

Around the margins of the present-day Wharton Basin, the Late Cretaceous is represented everywhere by detrital clays in its central parts these give way to true pelagic clays. Late Cretaceous limestones are found on Broken Ridge and limestones and dolomites at Site 217 on the northern Ninetyeast Ridge. Less is known about the western Indian Ocean sediments over this period. Early Late Cretaceous volcanogenic sediments on the Mozambique Ridge are overlain by calcareous sediments. In the western Somali Basin, Late Cretaceous sediments are inferred to be present on the basis of results from Site 242 and from seismic profiles in the region. Detrital clays relatively rich in pyrite and organic carbon, similar to those found in earlier times in the Wharton Basin, were encountered in the Mozambique Basin and off Madagascar. The clays off Madagascar are also rich in montmorillonite (Vallier and Kidd, 1977) implying nearby volcanism.

On the basis of the subsidence curves (Figs.4 and 5), the CCD in Campanian—Maastrichtian time was probably between 3000 and 4000 m. The deepest site at which in-situ carbonates were found was Site 211, then at a paleo-depth of about 3500 m. This allows us to map a "carbonate line" between sites on the 70-m.y.B.P. reconstruction (Fig.6), remembering to take account of depth fluctuations such as those that occur at the present-day near the

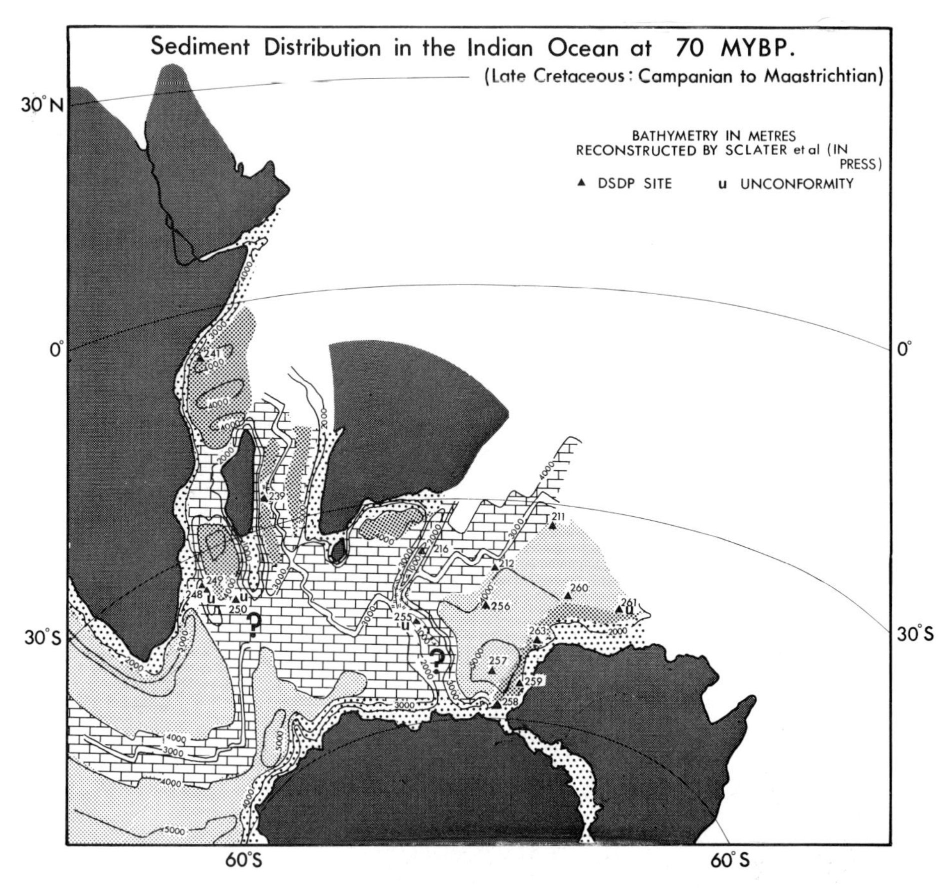

Fig.6. Sediment distribution in the Indian Ocean at 70 m.y.B.P. (Late Cretaceous): Campanian to Maastrichtian; symbols are as in Fig.2.

continents and in high latitudes. The resulting distribution of carbonate and non-carbonate sediments shows the developing basins generally to be sites for the accumulation of detrital clays, at high sedimentation rates and often under conditions of restricted circulation. Only sites in those parts of the Wharton Basin, which the reconstruction suggests were remote from land areas, contain typical pelagic clays. The reconstruction also shows an almost continuous topographic barrier between India and Australia/Antarctica formed by the Ninetyeast and Broken Ridges. The sedimentary evidence fits well with this. The southern end of Ninetyeast Ridge was the scene of volcanic activity as shown by the volcanic sediments and ash at Site 216, while the northern part, and presumably Broken Ridge and Naturaliste Plateau, were shallow platforms a few hundred meters deep on which pelagic carbonates accumulated. The placing of the spreading ridge close to Madagascar at this time is supported by the richness in montmorillonite of the clays in the developing Madagascar Basin (Site 239).

Paleogene sediments

Calcareous sediments of Paleocene age are found at sites on present-day ridges and plateaus, while basin sites contain pelagic clays over this interval. Off East Africa and west of India the Paleocene is a coarse terrigenous or silty detrital clay sequence. Volcanogenic sediments were sampled on the aseismic Ninetyeast and Owen Ridges. Elsewhere, such as along the Australian Margin, Broken Ridge and in the basins off Africa, the Paleocene is absent.

Early Eocene (Fig.7). Description of the Eocene sediment distribution is complicated because of gaps in the record at sites near the coasts and in the Mascarene Basin. The CCD was probably between 3000 and 4000 m (Figs.4 and 5). Unfossiliferous pelagic clays were encountered at the Wharton and Central Indian basin sites, which had already sunk below this level, and calcareous sediments occur at most other sites. Sediments at basin sites off Africa differ, however, in having a substantial silty terrigenous component and significant amounts of pyrite. Reference to the 53-m.y.B.P. reconstruction (Fig.7) shows that the sediment distributions can be explained

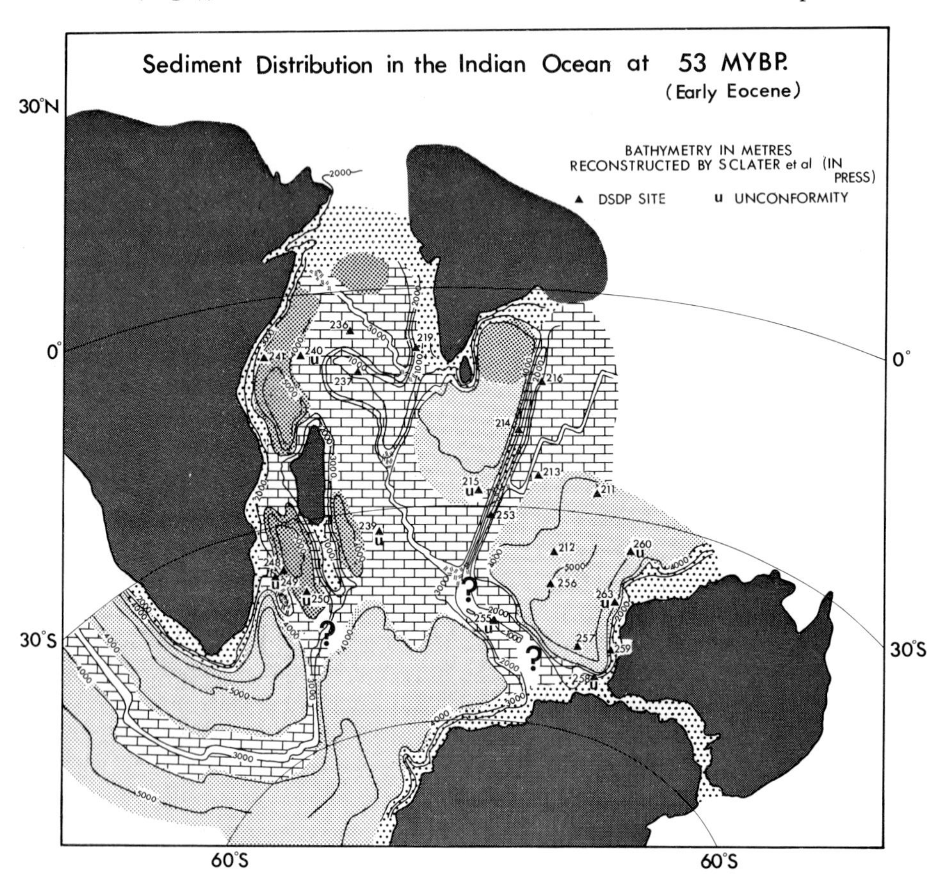

Fig.7. Sediment distribution in the Indian Ocean at 53 m.y.B.P. (Early Eocene); symbols are as in Fig.2.

by the degree of development of the basins. The extensive and relatively remote Wharton and Central Indian basins, with their pelagic clays, contrast with the smaller marginal basins in the west in which circulation may have been restricted.

The division of the Indian Ocean into eastern and western regions was apparently emphasised by the 30° latitudinal extent of the Ninetyeast Ridge from India, now athwart the Equator, to the point where it joined the Broken Ridge—Kerguelen Plateau—Naturaliste Plateau complex. The Wharton Basin, fringed to the north by carbonate platforms, associated with the Ninetyeast Ridge and a former mid-ocean ridge (Sites 216, 214 and 213), was steadily deepening, while remaining open to Tethys and the western Pacific to the east. As was clear from its widespread Eocene calcareous sediments, the western Indian Ocean remained shallow, dominated by the mid-ocean ridge and extensive shallow seas associated with the Chagos—Laccadive Ridge and Mascarene Plateau. The northwestern connection to Tethys was becoming constricted. Sedimentary evidence of continuing volcanic activity is found on the Ninetyeast and Mozambique Ridges and also on the Owen Ridge. The Early Eocene succession on Ninetyeast Ridge suggests deposition in very shallow water, which concurs with interpretations of the paleobathymetric curves and reconstruction. Similarly the presence of Cretaceous limestone on Broken Ridge, separated by a hiatus from Late Eocene littoral gravels, suggests that much of it was emergent as an island at this time.

Middle and Late Eocene patterns of sedimentation are essentially the same as in the Early Eocene, although the record is patchy and discontinuous. Two new developments occurred however; sometime in the Middle to Late Eocene Broken Ridge subsided below sea level, and over the same period occurred the first major influx of terrigenous material into the northern Arabian Sea, presumably as a result of the closure of the Indus Trough (Stonely, 1974) and uplift of the Himalayas.

Early Oligocene (Fig.8). The Early Oligocene is the most poorly represented part of the Paleogene record. Oligocene sediments are absent in almost all sites on the margins and also in some sites on the present-day ridges and plateaus, with the exception of those on the Mascarene and Chagos—Laccadive platforms. Despite its incompleteness, some observations may be made and we can infer sediment distributions to a degree, from underlying late Eocene or overlying late Oligocene deposits.

At Site 254 on the southern end of Ninetyeast Ridge, the Early Oligocene is represented by detrital volcanic sands and silts associated with lagoonal or littoral sediments, while other evidence of volcanic activity is afforded by Sites 234 and 238, presumably associated with Carlsberg Ridge volcanism. Other sites on Ninetyeast Ridge and Broken Ridge have patchy accumulations of calcareous ooze and the first Indus-derived sediments arrived in the southern Arabian Sea, Site 221 (Weser, 1974), during this period. It is also certain that the buildup of the Bengal Fan had begun by this time. Curray

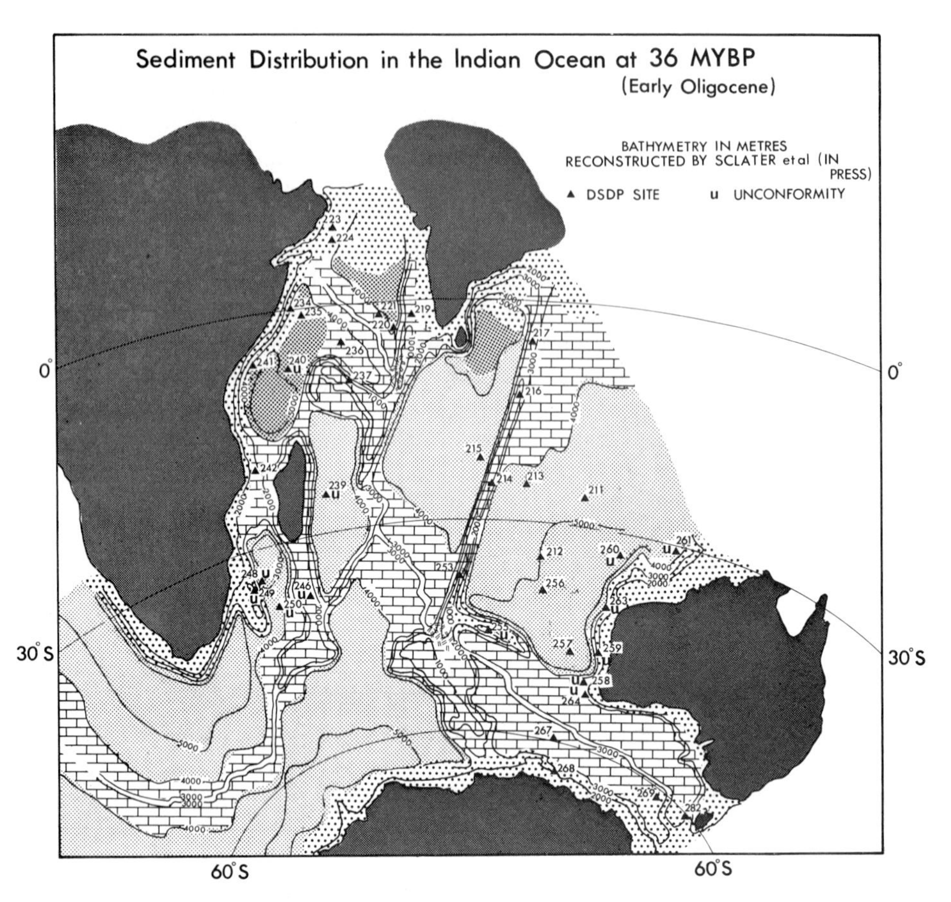

Fig.8. Sediment distribution in the Indian Ocean at 36 m.y.B.P. (Early Oligocene); symbols are as in Fig.2.

and Moore (1974) believe that much of this submarine fan's early sediments were deposited in deep basins off southeastern India, with the Godvari River as their probable source. The river, they postulate, began this buildup possibly as early as the Late Cretaceous. The Oligocene at all the present-day deep basin sites is represented only by thin unfossiliferous pelagic clays, which accumulated at very slow rates. They represent a dissolution facies that may contain considerable sedimentary hiatuses resulting from non-deposition (Davies et al., 1975).

Apart from widespread unconformities, the extensive input of terrigenous sediments on the northern submarine fans is the major new development in the Indian Ocean sediment distributions (Fig.8). The CCD at this time was probably at about 4000 m (Figs.4 and 5). Carbonates accumulated at deeper levels in the Somali and Arabian Basins, but these appear to have been displaced from shallower depths (Thiede, 1974). The Early Oligocene reconstruction suggests that the Indian Ocean now resembled its present-day configuration with three distinct regions: an almost totally enclosed northwestern region, divided by the Carlsberg Ridge and now with very limited access to Tethys; a central region, split by the inverted "Y" of the spreading ridges; and an eastern region, still with extensive communication to the western Pacific and dominated by a wide deep Wharton Basin. The eastern region by this time had additional communication with the Pacific, through a rapidly developing seaway to the southeast formed by the separation of Antarctica from Australia. The distribution of carbonate and non-carbonate sediments fits well the paleogeographic picture and the areas of pelagic clay are now much more extensive than at any previous period. This is due to the mature stage of development of most of the major basins: other types of deep sea clays are present only as distal sediments of the terrigenous accumulations in the north and west. The volcanic activity at the southern end of Ninetyeast Ridge appears to have been associated with the separation of the ridge from Kerguelen Plateau.

Neogene to Recent sediments

Since the Oligocene, the overall sediment distributions recognised from the drilled sequences essentially changed very little, although unconformities again present problems in determining distributions in the Early and Middle Miocene.

Terrigenous sedimentation increased in importance overall. Two notable developments in the Miocene were: (1) the first evidence of the presence of the Zambesi Fan in the Middle Miocene (Site 248) and its subsequent buildup towards the south (Site 250); and (2) the continued progradation of the East African margin, which reached into the eastern Somali Basin depositing detrital clays in the Lower Miocene; these clays surprisingly were not sustained into the Middle Miocene when carbonate sedimentation reasserted itself.

On the Bengal Fan, the influence of the Ganges—Brahmaputra system was by the Middle Miocene dominant (Site 218). Terrigenous sediments from this source, passed over the Nicobar Fan, and continued to accumulate east of the Ninetyeast Ridge (Site 211) at least until the beginning of the Pleistocene (Pimm, 1974). Active sedimentation ceased on the Nicobar Fan in the Middle Pleistocene. Miocene to Recent subduction of the Indian Plate under the Eurasian Plate created the Sunda Java Trench which, by the Middle Pleistocene, had developed sufficiently that sediments poured into its northern end so starving the Nicobar Fan. The associated development of the Indonesian volcanic arc caused an extensive buildup of volcanogenic sediments in this region in Pliocene to Recent times.

Continued movement of India northward and opening of the Gulf of Aden resulted in the destruction of eastern Tethys (Stonely, 1974; Blow and Hamilton, 1975). Progressive buildup of the Indus Fan meant that terrigenous sediments entirely dominate Miocene to Recent sequences north of the Carlsberg Ridge. In the Gulf of Aden drillsites, the Middle Miocene to Recent period is represented by mixed calcareous and detrital sediments which were invariably slumped. The latter contain a significant volcanogenic component, especially in the Lower Miocene. Siliceous sediments are reported from the Miocene and Pliocene in the northern Wharton Basin and Crozet Basin.

Carbonate accumulation, already increased by the development of extensive shallow regions on the flanks of the Southeast Indian Ridge, was generally enhanced after the Oligocene and particularly after the Miocene by drops in level of the CCD. This might explain why the Early Miocene clays of the eastern Somali Basin give way upwards to carbonate sediments. Spreading of the northern Wharton Basin sites to locations under high-productivity zones nearer the paleo-Equator explains the presence of siliceous sediments in the upper parts of their sequences. We infer that general oceanographic conditions, and in particular circulation, have been similar to the present for at least the last 10 m.y.

DISCUSSION

Having outlined the changing distributions of our five broad classes of sediment within the context of evolving paleobathymetry, we can now draw attention to some aspects which are of particular paleoenvironmental interest.

Carbonate sedimentation

The major development of calcareous oozes appears to be at the present day. This is not merely due to the ocean having progressively grown in size, but is strongly linked to its CCD level being deeper than at any time in the ocean's history.

The generally shallow level of the Mesozoic CCD is well established by Figs.4 and 5. Van Andel (1975) and Sclater et al. (1977) have analysed in detail the behaviour of the CCD level through time. Sclater et al. (1977) propose that the Indian Ocean CCD rose from approximately 3.5 km water depth in the Late Jurassic (approx. 140 m.y.B.P.) to less than 3 km water depth during the mid-Cretaceous (approx. 100 m.y.B.P.). After this they suggest it dropped to a level of around 4 km for all of the Paleocene, only to drop again to present-day levels of more than 4.5 km after the Miocene. Van Andel (1975) on the other hand, while agreeing on Mesozoic levels, suggested that there was a steady drop to around 4.4 km water depth in the Late Oligocene, which was followed by a steep rise to above 4 km in the Middle Miocene and then by an equally steep drop afterwards to present-day levels. Shallow Late Mesozoic CCD levels have been identified in the other oceans also (Berger and Von Rad, 1972). They have been variously linked to major transgressive events, providing extensive shelves as shallow-water carbonate sinks; to biological evolution, bringing about a scarcity of calcareous pelagic material or, more simply, of shell mineralogies that would be favourable for deep ocean preservation; to major changes in circulation, linked themselves to changing paleogeography; or to a combination of these factors. One reason for the disagreement between the two interpretations of Neogene CCD levels as outlined above is that the earlier publication was based partly on preliminary DSDP results. However, it is clear that the difficulty in differentiating reworked versus "in-situ" carbonates causes differing interpretations and thus further analysis of this aspect in the stored cores may be warranted. Secondly we must emphasise that the lack of Southern Indian Ocean drilling data is a severe handicap and, if we are to progress to an understanding of the mechanisms which cause CCD fluctuations, information about locations to the south is essential.

Clay sedimentation

Clay sediments probably had their maximum areal distribution around the early Miocene, when the deep basins were at their maximum extent prior to the most recent drop of the CCD to its present depth range. Since this time also, terrigenous sedimentation from the north and west has made major advances into the basins and siliceous oozes have accumulated in areas below the high productivity zones. Typical pelagic clays have been the major fine sediment type accumulating below the CCD throughout the evolution of the Indian Ocean. However, during the early stages of development of the individual deep basins, other types of deep sea clays have been deposited, usually at anomalously high sedimentation rates and often under conditions of restricted circulation. Similarly, distal deposits of submarine fans represent anomalous deep sea clays. At the present day, deep sea clays containing a terrigenous detrital component are accumulating at high sedimentation rates in the northern Mozambique and Mascarene basins. This is probably caused by strong currents, which sweep the narrow southeast African continental shelf, so moving sediment into these deep basins.

Siliceous sedimentation

Generally siliceous sediments are poorly represented in the Indian Ocean. It appears that, in the last 10 m.y., conditions have been suited to the development of zones of high siliceous plankton productivity and thus siliceous oozes in the deep basins.

Cherts are diagenetic evidence of strong input of siliceous fossils to dominantly calcareous sediments that were deposited above the CCD. None of the Indian Ocean cherts is post-Eocene in age and most beds or nodules that are encountered, are in Eocene or Paleocene sediments. Reconstructions of the paleobathymetry of these chert occurrences do indeed place them above the CCD and locate them either in coastal regions or close to an emergent ridge. Thus, we might infer that upwelling was the mechanism which caused these siliceous accumulations. For example, sites in the Arabian Sea and seismic profiling in the area suggest the presence of a well-defined, Eocene chert reflector (Whitmarsh et al., 1974). This may be linked to the establishment of gyral circulation in the Arabian Sea at this time, which brought about upwelling off the Arabian and Indian coasts. On the other hand, the distribution of Early Cretaceous cherts and sediments rich in radiolaria in the Wharton Basin drillsites and in land sections in northern Australia (Brown et al., 1968), when compared with the paleogeography of the time, suggests that these were linked to open ocean conditions and productivity within Tethys.

Terrigenous sedimentation

Terrigenous sediments represent the mechanical load of rivers and Garrels and Mckenzie (1971) have established the correlation between this load and mean continental elevation. The principal accumulations of terrigenous, clastic detrital, sediments lie along the northern and western margins of the Indian Ocean. The fact that comparatively little clastic detrital sediment is found in the eastern Indian Ocean is explained by Australia having been a lowlying landmass from early in the Mesozoic (King, 1962).

The four principal accumulations of terrigenous sediment are clearly associated with major tectonic events: the Indus and Bengal Fans with the first contact of the Indian subcontinent and Eurasia in the Eocene (Stonely, 1974), and their subsequent rapid buildup during the Oligocene and Miocene with the closure of the Indus Trough and uplift of the Himalayas; the progradation of the Somali Margin since the Early Cretaceous with the long history of tectonic activity in East Africa; and lastly the rapid growth of the Zambesi Fan from mid-Miocene times with the Neogene epeirogenic uplift and tilting of southern and central Africa (King, 1962). Girdley et al. (1974) have demonstrated that the history of tectonic events in the African hinterland closely correlates with the record of rates of sediment accumulation in the Somali and Mozambique basins.

Volcanogenic sedimentation

Kennett and Thunell (1975) analysed the distribution of Miocene to Recent volcanic ash from the lithological reports of 320 deep sea drilling sites and suggested that a global increase in explosive volcanism occurred in Quaternary times. Vallier and Kidd (1977), taking account of sediment accumulations made up of secondary as well as primary volcanogenic components, have demonstrated that input of volcanic material to the Indian Ocean has been fairly continuous, but nevertheless major pulses in volcanic activity can be recognised: in the Early and Late Cretaceous, the Eocene and the Oligocene, as well as in the Plio-Pleistocene. The Oligocene event is thought to have been even more significant than its distribution suggests, since much of our record of this period is missing due to hiatuses. The pulses are linked to tectonic events involved in the breakup of Gondwana and to changes in direction and rates of sea-floor spreading. These deductions are made with the exclusion of data from the aseismic ridges. The particular mode of formation of these ridges ensures the presence of volcanogenic sediments at the bases of sites drilled along them (Fleet and McKelvey, 1978, this issue).

Hiatuses and ocean circulation

The history of Indian Ocean sedimentation is characterised by periods of greatly reduced or total lack of sedimentation which had effect on an ocean-wide scale (Davies et al., 1974). Some of the gaps in the sedimentary record, at places around the margins and others on ridges can be attributed to tectonic activity, either directly (i.e. uplift and erosion) or indirectly (e.g. tectonic events, which interrupted sediment supply). However, a large number of the hiatuses occur in regions remote from the former land margins. Here the rate of sediment supply is largely a function of surface productivity and the rate of removal is determined by the strength of bottom currents and the tendency of cold deep water to dissolve calcite and silica. Thus the causes of these ocean-wide hiatuses should be linked to circulation effects.

The most conspicuous period of hiatuses in the Indian Ocean is the Early Oligocene. Hiatuses are found all around the margins, on shallow ridges and plateaus (with the notable exception of the Mascarene and Chagos—Laccadive platforms) and also in the deep basins (Figs.4 and 5), either as true unconformities or as dissolution facies resulting from reduced sedimentation rates or non-deposition.

At the present day, cold "aggressive" bottom water forms in the Antarctic shelf regions and spreads north into the southwest Indian Ocean, where it behaves as a western boundary undercurrent under which pre-Pleistocene sediments are frequently exposed. It also drifts northeastwards through the fracture zones in the Southeast Indian Ridge and passes, through the gap between Broken Ridge and the Naturaliste Plateau, into the Wharton Basin. This pattern of distribution is mirrored by the distribution of the Oligocene hiatus (Davies et al., 1974, p.18, fig.4) and this would suggest a causal relationship. Glaciation began in Antarctica near the end of the Eocene or the beginning of the Oligocene (Kennett et al., 1974). This must have produced copious amounts of cold "aggressive" bottom water which probably moved northwards into the Indian Ocean, thus establishing vigorous circulation and causing widespread erosion or non-deposition. Once deep circumpolar circulation was established in the Late Oligocene (Kennett et al., 1974), the intensity of Indian Ocean bottom water circulation probably decreased, so allowing sedimentation to resume in most places. A related series of events may be applicable to the other extensive Paleogene and Late Mesozoic hiatuses which formed as Antarctica moved southwards and its climate deteriorated; however, evidence of their regional distribution is too ambiguous at present to permit detailed analysis. Such speculation on the relationship of hiatuses to circulation effects is anyway severely compromised by the lack of southern high-latitude drillsites.

CONCLUSIONS

The 1974 deep sea drilling in the Indian Ocean has provided a lithostratigraphic record through which we can outline changing distributions of sediment types from the Jurassic to the present day. Subsidence curves and paleobathymetric reconstructions show how the patterns of sedimentation are controlled by the evolution of this relatively young ocean basin. The 53-m.y.B.P. and 70-m.y.B.P. reconstructions, considered by Sclater et al. (1977) to be speculative, fit well the mapped sediment distributions.

Mesozoic sediment distributions are characterised by relatively rapid accumulation of clays in restricted basins with calcareous and terrigenous sediments accumulating around their margins. In the Paleogene more open conditions prevailed but tectonic activity and changing bottom circulation resulted in a patchy and discontinuous record. The Neogene to Recent record is relatively continuous and extends through almost all of the drill-sites. However, it is clear that, by Neogene times, the major geographic features of the present ocean had already developed, as patterns of sedimentation since the Oligocene closely resemble those of the present day. The causes of perturbations in the CCD level and of oceanwide hiatuses are considered to be strongly linked to the history of changes in water mass circulation. We recognise, as do Kennett et al. (1976), the need for southern latitude drilling before we can expect to understand this history and its effects on planktonic faunal productivity.

ACKNOWLEDGEMENTS

We wish to emphasise that we have drawn heavily upon the observations of our colleagues, the scientific staffs of "Glomar Challenger", in preparing this paper. We hope we have given appropriate recognition in every case but apologise in advance for any omissions. We are greatly indebted to J.G. Sclater, D. Abbott and J. Thiede for allowing us use of their paleobathymetric reconstructions.

We thank R.B. Whitmarsh (IOS) for a critical review of the paper; P.M. Hunter and W.S. Blyth for drafting the figures and M. Neal for typing the manuscript.

The senior author acknowledges the close co-operation, help and guidance of O.E. Weser during his studies of the Indian Ocean cores stored at the DSDP West Coast Repository. Dr. M.N.A. Peterson (DSDP) and Professor H. Charnock (IOS) are acknowledged for arranging his secondment to DSDP Headquarters, La Jolla, which allowed these studies to be undertaken.

REFERENCES

Berger, W.H., 1972. Deep-sea carbonates: dissolution facies and age—depth constancy. Nature, 236: 392.

Berger, W.H. and Von Rad, U., 1972. Cretaceous and Cenozoic sediments from the Atlantic Ocean. In: Initial Reports of the Deep Sea Drilling Project, 14. U.S. Govt. Printing Office, Washington, D.C., pp.945—952.

Berger, W.H. and Winterer, E.L., 1974. Plate stratigraphy and the fluctuating carbonate line. In: K.J. Hsü and H.C. Jenkyns (Editors), Pelagic Sediments on Land and Under the Sea. Int. Assoc. Sedimentol., Spec. Publ., 1: 11—48.
Berggren, W.A., 1972. A Cenozoic time scale — some implications for regional geology and paleobiogeography. Lethaia, 5: 195—215.
Berggren, W.A. and Van Couvering, J.A., 1974. Biostratigraphy, geochronology and paleoclimatology of the last 15 million years in marine and continental sequences. Palaeogeogr. Palaeoclimatol. Palaeoecol., 16: 1—216.
Blow, R.A. and Hamilton, N., 1975. Paleomagnetic evidence from DSDP cores of northward drift of India. Nature, 257: 570—572.
Brown, D.A., Campbell, K.S.W. and Crook, K.A.W., 1968. The Geological Evolution of Australia and New Zealand, Pergamon, London, 409 pp.
Curray, J.R. and Moore, D.G., 1974. Sedimentary and tectonic processes in the Bengal deep-sea fan and geosyncline. In: C.A. Burk and C.L. Drake (Editors), Geology of Continental Margins. Springer, New York, N.Y., pp.617—628.
Davies, T.A. and Kidd, R.B., 1977. Sedimentation in the Indian Ocean through time. In: J.R. Heirtzler (Editor), Deep Sea Drilling in the Indian Ocean. JOIDES Spec. Publ., 1: in press.
Davies, T.A. and Luyendyk, B.P. et al., 1974. Initial Reports of the Deep Sea Drilling Project, 24. U.S. Govt. Printing Office, Washington, D.C., 1126 pp.
Davies, T.A., Weser, O.E., Luyendyk, B.P. and Kidd, R.B., 1975. Unconformities in the sediments of the Indian Ocean. Nature, 253 (5486): 15—19.
Fisher, R.L., Bunce, E.T. et al., 1974. Initial Reports of the Deep Sea Drilling Project. U.S. Govt. Printing Office, Washington, D.C., 1183 pp.
Fleet, A.J. van McKelvey, B.C., 1978. Eocene explosive submarine volcanism, Ninetyeast Ridge, Indian Ocean. Mar. Geol., 26: 73—97 (this issue).
Garrels, R. and MacKenzie, F.T., 1971. Evolution of the Sedimentary Rocks. Norton, New York, N.Y., 397 pp.
Girdley, W.A., Leclaire, L., Moore, C., Vallier, T.L. and White, S.M., 1974. Lithology summary, Leg 25 Deep Sea Drilling Project. In: E.S.W. Simpson, R. Schlich et al., Initial Reports of the Deep Sea Drilling Project, 25. U.S. Govt. Printing Office, Washington, D.C., pp.725—759.
Hayes, D.E., Frakes, L.A. et al., 1975. Initial Reports of the Deep Sea Drilling Project, 28. U.S. Govt. Printing Office, Washington, D.C., 1017 pp.
Jipa, D. and Kidd, R.B., 1974. Sedimentation of coarser-grained interbeds in the Arabian Sea and sedimentation processes of the Indus Cone. In: R.B. Whitmarsh, O.E. Weser et al., Initial Reports of the Deep Sea Drilling Project, 23. U.S. Govt. Printing Office, Washington, D.C., pp.521—525.
Kennett, J.P., Houtz, R.E., et al., 1974. Development of the Circum Antarctic Current. Science, 186: 144—147.
Kennett, J.P., Sclater, J.G. and Van Andel, T.H., 1976. For JOIDES: Southern Ocean targeted. Geotimes, 21 (3): 21—24.
Kennett, J.P. and Thunnell, R.C., 1975. Global synchronism and increased Quaternary explosive volcanism. Science, 187: 497—503.
Kennett, J.P. and Watkins, N.D., 1975. Deep-sea erosion and manganese nodule development in the Southeast Indian Ocean. Science, 188: 1011—1013.
Kidd, R.B. and Weser, O.E., 1974. A preliminary synthesis of sediments recovered by the Deep Sea Drilling Project in the Indian Ocean. Geol. Soc. Am. Annu. Meet., Miami, Fla., Abstract.
Kidd, R.B. and Weser, O.E., 1975. Sediment facies in the Indian Ocean. Int. Sedimentol. Congr., 9th, Nice, 5.
King, L., 1962. The Morphology of the Earth, Hafner, New York, N.Y., 699 pp.
Kolla, V., Be, A.W.H. and Biscaye, P.E., 1976. Calcium carbonate distribution in the surface sediment of the Indian Ocean. J. Geophys. Res., 81 (15): 2605—2116.

Luyendyk, B.P., 1974. Gondwanaland dispersal and the early formation of the Indian Ocean. In: T.A. Davies, B.P. Luyendyk et al., Initial Reports of the Deep Sea Drilling Project, 26. U.S. Govt. Printing Office, Washington, D.C., pp.945—952.
Luyendyk, B.P. and Davies, T.A., 1974. Results of DSDP Leg 26 and the geologic history of the southern Indian Ocean. In: T.A. Davies, B.P. Luyendyk et al., Initial Reports of the Deep Sea Drilling Project, 26. U.S. Govt. Printing Office, Washington, D.C., pp.909—943.
McKenzie, D. and Sclater, J.G., 1971. The evolution of the Indian Ocean since the Late Cretaceous. Geophys. J.R. Astron. Soc., 25: 437—528.
Pimm, A.C., 1974. Sedimentology and history of the Northeastern Indian Ocean from Late Cretaceous to Recent. In: C.C. von der Borch, J.G. Sclater et al., Initial Reports of the Deep Sea Drilling Project, 22. U.S. Govt. Printing Office, Washington, D.C., pp.77—803.
Ryan, W.B.F., Cita, M.B., Rawson, M.D., Burckle, L.H. and Saito, T., 1974. A paleomagnetic assignment of Neogene stage boundaries and the development of isochronous datum planes between the Mediterranean, the Pacific and Indian Oceans in order to investigate the response of the world ocean to the Mediterranean "salinity crisis". Riv. Ital. Paleontol., 80 (4): 631—688.
Sclater, J.G. and Fisher, R.L., 1974. The evolution of the east central Indian Ocean. Geol. Soc. Am., Bull., 85: 683—702.
Sclater, J.G., Anderson, R. and Bell, L., 1971. The elevation of ridges and the evolution of the central eastern Pacific. J. Geophys. Res., 76: 7888—7915.
Sclater, J.G., Abbott, D. and Thiede, J., 1977. Paleobathymetry and sediments of the Indian Ocean. In: J.R. Heirtzler (Editor), Deep Sea Drilling in the Indian Ocean. JOIDES Spec. Publ., 1: in press.
Simpson, E.S.W., Schlich et al., 1974. Initial Reports of the Deep Sea Drilling Project, 25. U.S. Govt. Printing Office, Washington, D.C., 883 pp.
Stonely, R., 1974. Evolution of the continental margins bounding a former Southern Tethys. In: C.A. Burk and C.L. Drake (Editors), Geology of Continental Margins. Springer, New York, N.Y., pp.889—906.
Thiede, J., 1974. Sediment coarse fractions from the western Indian Ocean and Gulf of Aden (Deep Sea Drilling Project, Leg 24). In: R.L. Fisher, E.T. Bunce et al., Initial Reports of the Deep Sea Drilling Project, 24. U.S. Govt. Printing Office, Washington, D.C., pp. 651—765.
Thiede, J. et al., 1975. Sediments on the Rio Grande Rise (S.W. Atlantic Ocean: subsidence of an aseismic ridge). Int. Sedimentol. Congr., 9th, Nice, 8.
Udintsev, G.B. (Editor), 1975. Geological—Geophysical Atlas of the Indian Ocean. Academy of Sciences of the USSR, Moscow, 152 pp.
Vallier, T.L. and Kidd, R.B., 1977. Volcanogenic sediments in the Indian Ocean. In: J.R. Heirtzler (Editor), Deep Sea Drilling in the Indian Ocean. JOIDES Spec. Publ., 1: in press.
Veevers, J.J., Heirtzler, J.R. et al., 1974. Initial Reports of the Deep Sea Drilling Project, 27. U.S. Govt. Printing Office, Washington, D.C., 1059 pp.
Van Andel, T.H., 1975. Mesozoic/Cenozoic calcite compensation depth and the global distribution of calcareous sediments. Earth Planet Sci. Lett., 26: 187—194.
Von der Borch, C.C., Sclater, J.G. et al., 1974. Initial Reports of the Deep Sea Drilling Project, 22. U.S. Govt. Printing Office, Washington, D.C., 889 pp.
Weser, O.E., 1974. Sedimentological aspects of strata encountered on Leg 23 in the Northern Arabian Sea. In: R.B. Whitmarsh, O.E. Weser et al., Initial Reports of the Deep Sea Drilling Project, 23. U.S. Govt. Printing Office, Washington, D.C., pp.503—519.
Whitmarsh, R.B., Weser, O.E. et al., 1974. Initial Reports of the Deep Sea Drilling Project, 23. U.S. Govt. Printing Office, Washington, D.C., 1179 pp.

Marine Geology, 26 (1978) 71—72

MESSINIAN EVAPORITES FROM THE MEDITERRANEAN AND RED SEAS

KENNETH J. HSÜ[1], PETER STOFFERS[2] and DAVID A. ROSS[3]

[1]*Geological Institute, ETH, Zurich (Switzerland)*
[2]*Laboratorium für Sedimentforschung, Heidelberg (F.R.G.)*
[3]*Woods Hole Oceanographic Institution, Woods Hole, Mass. (U.S.A.)*

(Received March 28, 1977)

ABSTRACT

Hsü, K.J., Stoffers, P. and Ross, D.A., 1978. Messinian evaporites from the Mediterranean and Red Seas. Mar. Geol., 26: 71—72.

A prominent subbottom reflector (M- in the Mediterranean and S- in the Red Sea) has been identified by deep sea drilling as the top of an Upper Miocene evaporite formation. The Mediterranean evaporite minerals are dolomite, gypsum, anhydrite, halite, polyhalite, kainite, sulphoborite, possibly bishofite and lunebergite, a borate and phosphate mineral. The Red Sea evaporite minerals include dolomite, magnesite, gypsum, anhydrite, halite, polyhalite, and possibly tachyhydrite. Analcite and length-slow chalcedony are present locally in both regions, probably as diagenetic minerals.

The sedimentary structures of the evaporites from both regions indicate a genesis in sabkha environments. The bromine profiles of the halites are also typical of deposition in shallow brine-pools. On the other hand, the sediments immediately below and above the evaporite in the Mediterranean, and those immediately above the evaporite in the Red Sea are typically hemipelagic marls, containing deep-water benthonic fossils. We believe, therefore, that the evaporites were deposited in pre-existing deep basins, and that the evaporites, at least those from the intervals penetrated by deep sea drilling, were deposited after the basins were partially or wholly desiccated.

Isotopic analyses of the carbonates and of the interstitial waters from the evaporites suggest considerable freshwater influx into the desiccating basins. That the Messinian basins were host to fresh or brackish water lakes is further proven by the occurrence of a Paratethys ostracod-fauna and an epiphytic diatom-flora in the upper levels of the Mediterranean evaporites. While the Red Sea might have received its freshwater supply from African rivers, the eastern Mediterranean *lac mer* with its *Cyprideis pannonica*, might have been a part of the Paratethys, which was a great brackish water body extending from Vienna Basin to Aral Sea.

The stratigraphy of the evaporites under the Red Sea is remarkably similar to that under central Ionian abyssal plain (Site 374). A dark gray, dolomitic marl (or mud) unit is present between an anhydrite—halite unit of Late Messinian and a normal marine hemipelagic marl of Pliocene age. The dolomitic sediment is largely unfossiliferous except for some nannofossils (latest Miocene or earliest Pliocene) and it may represent a more euxinic, or more basinal equivalent of Upper Messinian *Cyprideis*-bearing beds elsewhere (e.g. Florence Ridge, Sites 375—376).

The simultaneous occurrence of an evaporite by desiccation during the latest Miocene and the simultaneous submergence under normal marine waters during the earliest Pliocene suggest that the Mediterranean and the Red Sea Basins were parts of one and the same hydrographic system. As the Mediterranean is demonstrably an Atlantic province during the Mio-Pliocene, we suspect that the Red Sea was not connected with the Indian Ocean when its evaporites were being formed. The Strait of Bab el Mandeb was opened during a more recent episode of seafloor spreading. This event, together with an isolation from the Mediterranean by the Isthmus of Suez, rendered the Red Sea an appendix of the Indian Ocean.

Marine Geology, 26 (1978) 73—97

EOCENE EXPLOSIVE SUBMARINE VOLCANISM, NINETYEAST RIDGE, INDIAN OCEAN

A.J. FLEET and B.C. McKELVEY

Department of Earth Sciences, The Open University, Milton Keynes, Buckinghamshire (Great Britain)
Department of Geology, University of New England, Armidale, N.S.W. (Australia)

(Received March 28, 1977)

ABSTRACT

Fleet, A.J. and McKelvey, B.C., 1978. Eocene explosive submarine volcanism, Ninetyeast Ridge, Indian Ocean. Mar. Geol., 26: 73—97.

A basaltic sequence of Eocene submarine-erupted pyroclastic sediments totals at least 388 m at DSDP Site 253 on the Ninetyeast Ridge. These fossiliferous hyaloclastic sediments have been erupted and fragmented by explosive volcanism (hydroexplosions) in shallow water. The occurrence of interbedded basaltic ash-fall tuffs within the younger horizons of the hyaloclastic sequence marks the emergence of some Ninetyeast Ridge volcanic vents above sea level. Considerable textural variation allows subdivision of the sequence into six informal lithostratigraphic units. Hydrothermal and diagenetic alteration has caused the complete replacement of all original glass by smectites, and the introduction of abundant zeolite and calcite cements.

The major and trace element contents of the hyaloclastites vary due to the alteration, and the admixture of biogenous calcite. On a calcium carbonate-free basis systematic variations are recognisable. Mg, Ni, Cr and Cu are enriched, and Li and Zn depleted in the three older units relative to the younger three. The chemical variability is reflected by the development of saponite in the older part of the sequence and montmorillonite in the younger; and by the presence of a quartz-normative basalt flow occurring in Unit II, in contrast to the Mg-rich highly olivine-normative basalt at the base of the sequence. The younger and older parts of the sequence therefore appear to have been derived from magmas of different chemistry.

The sequence, like other basaltic rocks recovered from the Ninetyeast Ridge, is enriched in the light relative to the heavy rare earth elements (REE) although the REE contents vary unsystematically with depth, probably because of the high-temperature subaqueous alteration and the presence of biogenous calcite. This REE data indicates that the Ninetyeast Ridge volcanism was different from that which produces mid-ocean ridge basalts.

FORMATION OF THE HYALOCLASTITE SEQUENCE

Introduction

During Leg 26 of the Deep Sea Drilling Project, Site 253 situated on the Ninetyeast Ridge penetrated a 388-m sequence of Middle Eocene marine basaltic hyaloclastites interbedded with subordinate terrigenous (basaltic source) sediments. The hyaloclastic sequence contains some interbedded basaltic ash-fall tuffs in the younger horizons, and is overlain with apparent conformity by 153 m of younger Eocene to Quaternary chalks and nannofossil oozes. In the Initial Reports of Leg 26 (Davies et al., 1974), the petrography of the hyaloclastites was described in some detail and the basic geochemical parameters of the sequence were outlined (McKelvey and Fleet, 1974, pp.553—565). However no real attempt was made to interpret the physical nature of the volcanism that produced the hyaloclastic sediments. Neither was the presence of ash-fall tuffs (i.e. subaerial eruptives) recognized. It was assumed the hyaloclastites were the result of very hot lava (1000°C) encountering seawater and that the consequent drastic tempering had caused fragmentation. No decision was made as to whether the volcanism was submarine, or alternatively a subaerial event with the lava flowing down into the sea and then fragmenting. We now feel the latter is definitely not the case. Furthermore although the different varieties of hyaloclastites were described, the significance of some of the overall features of the sequence was not fully considered. Two examples are the scarcity of associated lava flows, and the complete lack of basalt pillows and pillow breccias.

In this paper the mode of formation of the hyaloclastite sequence is reconsidered. This leads to some appreciation of the nature of the parent eruptions, and in consequence also some detail of the volcanic evolution and palaeogeographic setting of the Ninetyeast Ridge in Eocene times. B.C. McKelvey is responsible for the views expressed in this section of the paper. Other DSDP cores recovered previously from the Ninetyeast Ridge (e.g. DSDP 214, 216) have produced palaeogeographic data somewhat different from that indicated by Site 253. The documentation and integration of all such palaeogeographic data is essential for the substantiation and refinement of dynamic tectonic models proposed for the origin of the enigmatic Ninetyeast Ridge.

In the second part of this paper the major geochemical trends apparent in the hyaloclastite sequence are interpreted: (1) in relationship to the middle Eocene volcanism and the subsequent diagenetic history of the sequence; and (2) in relationship to other Ninetyeast Ridge basalts and also Indian Ocean basalts in general. A.J. Fleet is responsible for these data and their interpretation.

Hyaloclastite sequence

The main features (Figs.1—4) of the sequence are outlined below. For details of the petrography the reader is referred to McKelvey and Fleet (1974, pp.554—560).

(1) The 388-m sequence is composed very largely of hyaloclastite sediments, of basaltic derivation. With the exception of 30 cm and 16 cm of vesicular or scoriaceous basalt recovered at 219 m and 228 m respectively, basaltic lavas are absent. (These two thin basalt horizons could alternatively be sills.) Furthermore there is a complete absence of basalt pillows or their

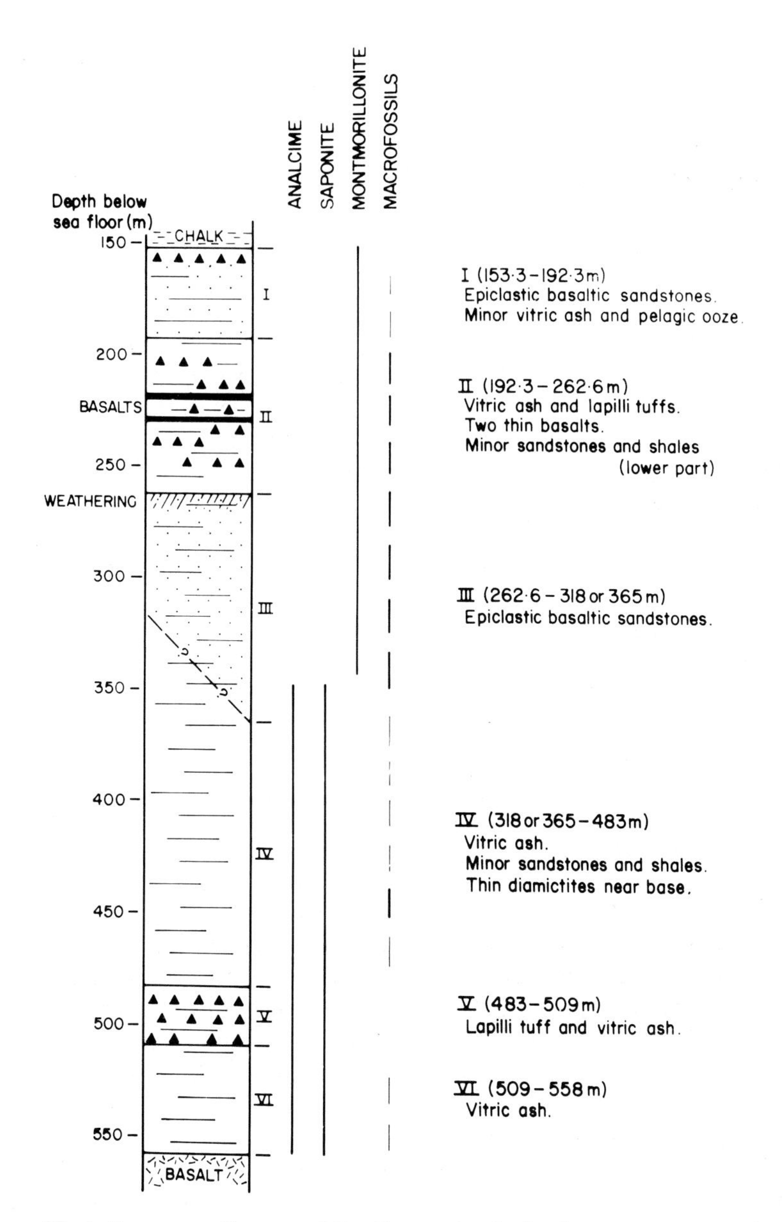

Fig.1. Summary diagram of the Eocene hyaloclastic strata penetrated at DSDP Site 253. Textural variation allows recognition of six lithostratigraphic units that make up the 388-m sequence. Units I and III have been stippled to contrast their largely epiclastic nature with the predominantly hyaloclastic Units II, IV, V and VI. The horizontal lines throughout the diagram are to emphasize the well-stratified nature of the sequence. Relative abundance of hyaloclastic lapilli tuffs is indicated by the black triangles. The portrayed thicknesses of the two basalts in Unit II is much exaggerated, and the thickness of the basal basalt is not known. The position of the boundary between Units III and IV is uncertain, because of poor core recovery. The presence of basaltic subaerially erupted ash-fall tuffs in Unit II has not been indicated on the diagram.

fragmented derivatives, pillow breccias (Carlisle, 1963).

(2) The sequence is not uniform and lithological variation allows recognition of six (informal) units.

(3) These units are internally highly stratiform. Bedding thicknesses throughout range from about 2.5 m down to less than 1 cm. In Unit II bedding contacts are sharp and may even be characterized by small-scale scour relationships. Some small (<1.5 cm) cross-strata and also load casts are present. In the older units stratification is less distinct and many boundaries are gradational. Sedimentary structures are not apparent.

(4) Units II, IV, V and VI consist almost wholly of primary hyaloclastites. These are either vitric ashes (average grain size <1 mm) or, less commonly, coarse lapilli tuffs (average grain size >4 mm). However, poorly sorted admixtures of these two petrographic types are not uncommon. Primary (detrital) matrices are not present, the sediment frameworks being cemented by zeolites, calcite and smectites. The lapilli tuffs make up the bulk of Unit V (483—509 m) and are common in Unit II. Elsewhere all other primary hyaloclastite horizons are composed largely of vitric ashes. It is the lapilli tuffs, some of which contain fragments up to 5.5 cm, which we consider most important in indicating both the nature of volcanism and the mechanism of hyaloclastite formation.

(5) The original glass of both the vitric ashes and the lapilli tuffs was generally devoid of pyrogenic crystals. In Unit II all glass fragments have been altered to montmorillonite (? via palagonite) and the sediment frameworks are cemented largely by calcite. Within Units IV, V and VI saponite similarly completely replaces the formerly glassy fragments and this framework is cemented by calcite and analcime (Figs.2, 3). Hemicrystalline and often scoriaceous xenoliths are common in the lapilli tuffs (e.g. McKelvey and Fleet, 1974, p.558, Fig.8).

(6) Units II, IV, V and VI represent virtually continuous hyaloclastite sedimentation, rapid enough to suppress (by dilution) recognisable intervals of non-volcanic pelagic sediments. In thin-section the vitric ashes do reveal a very minor microfossil contribution. With the exception of a 40-cm thick diamictite at the base of Unit IV, none of these rocks appear to be resedimented mass-movement deposits.

(7) In marked contrast, Units I and III are best regarded as composed of epiclastic sediments of basaltic provenance. These sediments appear to be an admixture of reworked pyroclastic and terrigenous sediments derived from a basaltic terrain. Several thin intervals (<6 cm) of vitric ash (and nannofossil ooze) have been recorded in Unit I. However in the main the two units reflect volcanic quiescence. This is further emphasized by the 1-m thick submarine-weathered interval at the top of Unit III.

(8) Basaltic ash-fall tuffs (i.e. subaerial eruptives) occur interbedded with the marine hyaloclastites in Unit II. (The limited sampling and thin-sectioning, due to the lack of sufficient induration within the Site 253 sequence does not allow the relative proportion and distribution of these sediments to be accurately determined.) We had previously regarded the ash-fall pyroclastics

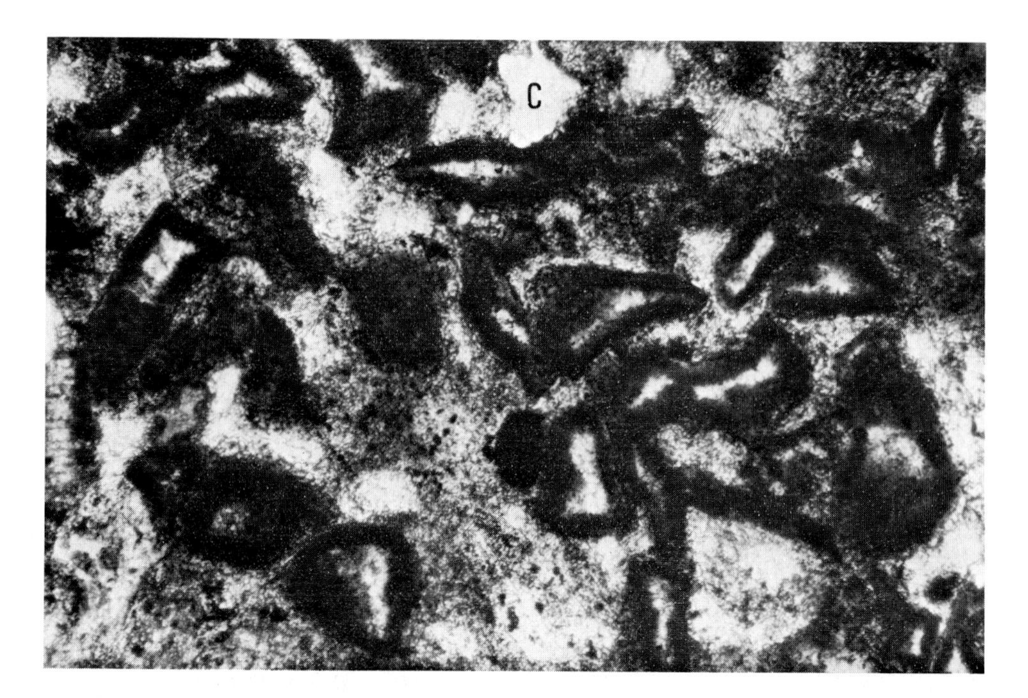

Fig. 2. Hyaloclastic vitric ash (DSDP Sample 253-40-2, 71—78). Tabular to triangular or irregular shaped angular grains of basaltic glass (shards) showing peripheral replacement to fibrous saponite. Interstitial to the shards is a secondary saponite matrix containing lighter coloured patches of analcime. Also present is a fragment of organic calcite (*C*). Plain light. Field width 2.2 mm.

as hyaloclastic sediments. The ash-fall tuffs consist of delicate basaltic pumice and reticulite fragments now replaced by montmorillonite, dispersed in a montmorillonite cement (Fig.4). In every way the intricate morphologies of the formerly vitric particles contrast with the much simpler splinter like shards of the hyaloclastic vitric ashes, and the blocky angular components of the lapilli tuffs.

(9) Marine microfossils and macrofossils are present throughout much of the sequence but their respective distributions vary. Microfossils (including nannofossils) are present in all units except the lapilli tuffs of Unit V and similar horizons of Unit II.

(10) Macrofossil detritus is absent from Unit V and from the lapilli tuff horizons of Unit II. It is rare in Unit I. The macrofossils are most abundant and best preserved in Units II and III. This feature taken in conjunction with the nature of stratification throughout the sequence (see 3, above) suggests a decrease in bathymetry from Unit VI up to and including Unit II. As such this part of the sequence reflects (relative) regressive hyaloclastic and epiclastic sedimentation. Comparison of Unit II with Unit I similarly suggests an upwards increase in bathymetry, and hence relative transgression.

(11) Interpretation of the transition from the hyaloclastite sequence to the overlying chalks and nannofossils oozes is not straightforward, and is

Fig.3. Hyaloclastic lapilli tuff (DSDP Sample 253-52-4, 5—16). Irregular to blocky shaped angular fragments of formerly basaltic glass now replaced by saponite. Interstitial cements are calcite and analcime. Lighter coloured framework fragments are basaltic xenoliths. Attitude of bedding is approximately perpendicular to length of core. Note the slightly inclined plane parallel fabric. Length of specimen 12 cm.

complicated by poor core recovery. The youngest preserved horizon of the otherwise essentially epiclastic Unit I is a thin (<12 cm) very coarse hyaloclastite breccia overlain by chalks and oozes devoid of any volcanic contribution. Either cessation of the final phase of volcanism was remarkably abrupt, or else a disconformity is present. Beneath the cored sequence the basal basalt at 558 m would appear on geophysical evidence not to be true basement, being probably underlain by more hyaloclastites and lava flows (Davies et al., 1974, p.167).

Discussion

Rittman (1962) introduced the term hyaloclastite to specify certain basaltic breccias formerly included under the very general and essentially petrologic term of palagonite tuff. Since the early usage of the term palagonite tuff by Italian workers (e.g. Sartorius v. Walterhausen, 1846, 1853; Platania, 1891, 1902—03; cited in Cucuzza-Silvestri, 1963) it has been gradually extended to refer to many different types of clastic basaltic units of diverse origins. Rittman (op. cit.) therefore proposed the term hyaloclastite for only those basaltic breccias whose fragmentary nature is a result of their mode of eruption. The term has now become widely used and most authors attribute the formation of hyaloclastites to either the brecciation of the outer crusts of pillowing lavas (i.e. the desquamation process of Rittman, 1958, 1962), or

Fig.4. Basaltic subaerially erupted ash-fall tuff. (DSDP sample 253-22-1; 101—103). Highly vesicular pieces of pumice and reticulite set in a brown cryptocrystalline smectite (montmorillonitic) cement. The pumice and reticulite varies from barely discernible highly transparent varieties to turbid and even opaque fragments. A crystal of fresh clinopyroxene (*p*) is present. The soft smectite is heavily scratched by grinding powder. Plain light. Field width 3.4 mm.

to the fragmentation of submarine lavas through sudden drastic water tempering (i.e. the granulation process of Bonatti, 1970; the quenching brecciation process of Cucuzza-Silvestri, 1963). Tazieff (1972, pp.472—473) in appraising the general validity of these processes accepts the former, but considers the latter two incapable of producing thick voluminous sequences of hyaloclastites.

None of these three processes appear suitable to account for the hyaloclastites encountered at Site 253. Here the absence of pillow lavas, solitary pillows and even pillow breccias precludes Rittman's desquamation process. Similarly the granulation process of Bonatti cannot be entertained inasfar as associated lava flows are virtually lacking. Admittedly the vitric ashes could have been produced in such a manner and then been carried far from the site of their parent lavas, either by ocean currents or alternatively by convecting seawater heated by submarine basaltic eruptions, a process suggested by Scheidegger (1972, p.401). However the coarse lapilli tuffs could not have been transported in this manner for such currents would not have been sufficiently competent. Finally the stratiform nature of the Site 253 hyaloclastics, and the lack of coarse massive intrusional breccias precludes any process similar to the quenching brecciation of Cucuzza-Silvestri.

Tazieff (1972, pp.473—474) maintains that the comminution of basalt lava to form any thick hyaloclastite sequence requires first the explosive eruption of basalt out of the sea floor from a submarine vent, and then the subsequent pulverization of the fragmented lava by closely successive phreatic (steam-generated) explosions. He maintains the process continues until "almost all the thermal energy contained in the original lava body . . . be converted into kinetic energy". As examples he cites the well-documented eruptions of Surtsey in Iceland and Capelinhos in the Azores. The violently explosive episodes characterizing the submarine eruptions of both volcanoes have been termed hydroexplosions by some authors. No attempt is made by Tazieff to describe in detail the mechanism of explosive fragmentation, that author simply maintaining that following the initial explosive eruption subsequent comminution by phreatic explosions results from steam entrapment in vesicles and umbrella-shaped bombs.

McBirney (1963) suggested that submarine explosive eruptions (i.e. hydroexplosions) capable of fragmenting magma result from the violent expansion (vesiculation) of original magmatic water. The fragmentation results when the coalescence of adjacent expanding vesicles causes loss of coherence of the magma, in the manner envisaged by Verhoogen (1951). Where the magmatic water content was approximately 0.1% and the magma temperature 1000° C, he estimated that such a process can occur down to depths of 500 m. A water content of 0.1% is high for basaltic rocks and so the figure of 500 m can be regarded as maximum value (pp.462—464).

Two recent publications (Colgate and Sigurgeirsson, 1973; Peckover et al., 1973) have drawn attention to an alternative hydroexplosion mechanism that also results in the fragmentation of magma erupted by submarine volcanoes in shallow-marine settings. These hydroexplosions are similar to the industrial hazard known as "fuel—coolant interactions" or "vapour explosions", an explanation of which has been proposed recently in some detail by Buchanan and Dullforce (1973). Fuel—coolant interactions result when "two molten materials, one (the fuel) much hotter than the other (the coolant) suddenly come in contact in such a manner that rapid heat transfer takes place to the cooler material which vaporises and expands explosively; physical rather than chemical processes are involved, and a large amount of thermal energy is transferred from the fuel to the coolant" (Peckover et al., 1973, p.307). As examples of such hydroexplosions these authors also cite the eruptions of Surtsey and Capelinhos, drawing attention to the black "cypressoid-shaped" jets of tephra that at Surtsey rose up to 150 m out of the sea, accompanied by clouds of steam.

We now consider the lapilli tuff grade hyaloclastites at Site 253 to have also been formed by volcanic hydroexplosions. The two-part sequence of fragmentation envisaged by Tazieff for such events suggests to us that perhaps both hydroexplosion mechanisms (i.e. those of McBirney; and of Buchanan and Dullforce) are operative in submarine eruptions of this nature. For a fuel—coolant interaction (hydroexplosion) to occur rapid mixing of seawater and magma are first necessary (Peckover et al., 1973). This could well be accomplished by a McBirney-type of hydroexplosion first erupting the fragmented magma out of the sea floor. With seawater and magma thus mixed, the rapidly successive fuel—coolant interactions outlined by Buchanan and Dullforce would follow, hurling jets of comminuted tephra similar to the Surtseyan "cypressoid clouds" out of the sea into the air. This ejecta would then fall back into the sea to be deposited as the lapilli tuff grade hyaloclastites. The admixture of highly vesicular and vesicle-poor fragments within the lapilli tuffs (e.g. McKelvey and Fleet, 1974, fig.9, p.558) may well indicate the operation of both McBirney type hydroexplosions and fuel—coolant interaction hydroexplosions. Only in the former is fragmentation brought about by vesiculation. Hyaloclastic fragments from both Surtsey and Capelinhos exhibit low vesicularity (Heiken, 1972, p.1977, fig.10).

The stratiform nature and coarse grain size of the hyaloclastic lapilli tuffs and the relative abundance of "basement"-derived volcanic xenoliths are then accounted for by a hydroexplosion mechanism, as also is the absence of pillow lavas and pillow breccias. Furthermore the moderate degree of sorting brought about by the lack of an interstitial fine fraction reflects settling of the comminuted detritus through a water column. Post-depositional fragmentation or spalling occurred as the still hot fragments making up the lapilli tuffs continued to cool after deposition (McKelvey and Fleet, 1974, p.559), a process discussed at some length by Carlisle (1963, p.61).

The fine fractions of the hyaloclastites separated by settling through a water column have been subsequently deposited as vitric ashes. Conceivably these fine-grained sediments may have undergone considerable lateral transport, being carried by marine currents. The high proportion of vitric ash to lapilli tuff does suggest Site 253 to have been distal rather than immediately proximal to the volcanic vent(s).

The coarse hyaloclastic breccia capping the sequence at the top of Unit I (Fig.1) shows submarine hydroexplosion activity continued until the cessation of volcanism. However the presence of the interbedded basaltic ash-fall tuffs in Unit II (i.e. above approximately 262 m, Fig.1) clearly evidences subaerial pyroclastic activity, and so marks the emergence of some of the Ninetyeast Ridge volcanic vents above sea level.

Bathymetry. Palaeontologic and volcanic data suggesting the bathymetry of the submarine eruption(s) that produced the Site 253 lapilli tuffs are broadly compatible. The palaeontologic evidence is the more precise and consists of the large benthonic foraminifera of the family *Discocyclina* in vitric ashes closely associated with the lapilli tuffs, indicating a maximum depositional depth (Davies et al., 1974, p.167) and therefore a (nearby) eruptive depth of about 150 m. As mentioned previously there is little evidence of mass movement and re-sedimentation within the hyaloclastite sequence.

Basaltic hydroexplosions resulting from the mechanism envisaged by McBirney can occur at depths down to 500 m. On the other hand if accepting the hydroexplosion mechanism proposed by Buchanan and Dullforce (1973) the maximum depth for fuel—coolant interactions varies between 130 and

700 m, according to the precise nature of the interaction (Peckover et al., 1973, p.308). Accepting the greater precision of the palaeontologic data, it would then appear reasonable to assume a maximum depth of less than about 150 m for the submarine volcanism that produced the lapilli tuffs, and the hyaloclastite sequence as a whole.

GEOCHEMISTRY OF THE HYALOCLASTITE SEQUENCE

Analytical techniques

Twenty-nine selected hyaloclastic and epiclastic samples were analysed by atomic absorption spectrophotometry for some major and trace elements. After the samples were oven-dried at 105°—110°C or freeze-dried, duplicate solutions of each were made up as described in McKelvey and Fleet (1974, p.560). The resulting solutions were analysed using a Perkin-Elmer 303 spectrophotometer, with a Perkin-Elmer HGA 2000 heated graphite atomiser being used for the analysis of Ba, Cr, Co and Mo. Detection limits for the instrument used were determined experimentally to be 1 ppm for Al, and Ti; 0.05 ppm for Fe, Ba, and Ni; 0.04 ppm for Sr; 0.02 ppm for Ca, and Cu; 0.01 ppm for Mn, Mg, Na, K, Li, Cr, Co, and Zn; and 0.005 ppm for Mo. The precision between duplicate solutions was generally better than 10%, and in the main was better than 5% for the major elements. Aliquots of the US National Bureau of Standards reference material 1b (argillaceous limestone), and the British Chemical Standards 267 and 269 (silica brick and firebrick, respectively) were analysed to determine accuracy. The determined values in nearly all cases were better than within 15% and were generally within 10% of the recommended values for these standards.

The selective fraction dissolution technique of Chester and Hughes (1967) was used in determining the $CaCO_3$ contents of all the samples, and the fractionation of Mn, Ni and Cu in some samples. The $CaCO_3$ contents of the samples were derived by leaching duplicate aliquots of the samples with 25% (v/v) acetic acid, and measuring the Ca contents of the resulting solution by atomic absorption spectrophotometry. Because surface-adsorbed elements are also released by this treatment (Chester and Hughes, 1967), and because the $MgCO_3$ content of marine biogenous material is low compared with the $CaCO_3$ content (e.g. Milliman, 1974), no attempt was made to measure the $MgCO_3$ contents of the samples. The fractionation of Mn, Ni and Cu in the selected samples was determined by leaching duplicate aliquots of each sample according to the technique of Chester and Hughes (1967). 25% (v/v) acetic acid released adsorbed elements and dissolved carbonates, except dolomite, and some Fe phases, whilst the mixed acid-reducing agent (1*M* hydroxylamine hydrochloride and 25% (v/v) acetic acid) dissolved Fe—Mn minerals. In addition 50% hot concentrated hydrochloric acid was assumed to dissolve all but the most resistant minerals (Cronan and Garrett, 1973).

Results

The analytical results for the twenty-nine samples are given in Tables I and II, together with the depths below the sea floor and lithological units from which the samples came. Some of the results are verified and revised values for the preliminary analyses given in McKelvey and Fleet (1974, pp.561—562). The results of selective dissolution are given in Table VII.

Discussion

Elemental variations with depth. The samples all consist of minerals resulting from the effects on basaltic material of varying degrees of alteration and erosion. They also contain variable micro- and macro-faunas, most of which are calcareous. As a result the compositions of these samples will reflect three factors to varying degrees: (1) the original type of parent basaltic material; (2) chemical exchanges which occurred during alteration and erosion of these materials; and (3) the admixture of biogenous phases, particularly the addition of $CaCO_3$.

These three factors must be taken into account when considering the analytical results. There is no way of quantifying the extent of the chemical changes due to alteration reactions; therefore evidence of the original composition of the parent igneous rock must be sought either in the gross chemical contents of the sediments, or in the nature of those elements which are thought to be little affected by alteration. The REE (rare earth elements) have mainly been used in furtherance of the latter approach (Fleet et al., 1976).

The problems arising from the admixture of biogenous $CaCO_3$ may be minimized by recalculating the results on a calcium carbonate free basis. This is not entirely satisfactory because inorganic calcite, formed as a result of alteration, is also present in the samples, and is not readily distinguishable from the biogenous variety by chemical analysis or X-ray diffractometry. In general, the inorganically produced $CaCO_3$ present in the sequence has been considered to be insignificant quantitatively.

For these reasons the data given in Tables I and II are shown recalculated on a calcium carbonate-free basis in Tables III and IV. These recalculated elemental contents may be divided into four groups:

(1) Those elements which have higher (Mg, Ni, Cr, and Cu) or lower (Li, Zn, and Mo?) abundances in the older three lithological units (IV, V, and VI) than in the upper units (I, II, and III).

(2) Those elements which vary from unit to unit (Ti, Fe?).

(3) Those elements which vary unsystematically with depth (Ca, Na, K, Al, Ba and Sr), or do not vary significantly (Co).

(4) Manganese, which decreases in abundance with depth.

(1) The magnesium contents of Units IV, V and VI are generally higher than those of the upper three units (Tables V and VI; Fig.5). This was indicated by the analyses reported in the Leg 26 Initial Reports (McKelvey and Fleet, 1974) which together with further data are presented here (Tables

TABLE I

DSDP Site 253 — Middle Eocene volcanogenic sequence — Major element analyses (%)

Core section	Interval (cm)	Depth below sea floor (m)	Ti	Al	Fe	Mn	Mg	Ca	Na	K	$CaCO_3$	Lithological units (McKelvey and Fleet, 1974)
17-2	60—62	154	1.93	6.47	10.92	1.08	2.22	1.26	2.77	2.52	1.95	I
18-1	60—62	162	0.99	3.07	4.62	0.22	2.03	1.41	1.65	0.88	47.10	I
19-2	9—11	172	1.21	3.70	6.69	0.16	2.04	3.21	1.73	1.49	34.39	I
20-2	90—92	182	0.90	2.63	3.88	0.12	1.99	4.15	1.68	0.56	47.77	I
21-2	58—60	192	0.07	5.35	10.64	0.02	5.02	1.96	2.55	0.84	1.66	I
21-4	112—114	195	1.41	6.36	7.37	0.06	3.60	1.91	2.30	1.34	9.16	II
23-2	86—88	211	0.74	6.51	5.23	0.10	3.14	2.69	2.02	1.44	18.35	II
26-2	23—25	239	1.15	8.20	6.39	0.04	4.60	3.90	2.49	0.84	1.81	II
28-2	18—19	252	0.88	6.42	4.65	0.39	3.84	2.15	1.89	0.68	18.18	II
30-1	120—122	271	0.62	2.78	6.48	0.10	2.20	1.98	1.01	1.51	42.23	III
31-2	118—120	281	1.87	6.54	7.82	0.07	4.18	2.06	2.16	1.03	6.54	III
36-1	93—95	327	0.43	7.31	4.26	0.01	3.97	1.87	2.54	0.70	1.25	III or IV
38-1	101—103	347	2.22	5.56	8.10	0.08	3.23	1.76	2.33	0.96	14.93	III or IV
40-2	30—32	366	1.45	5.44	7.53	0.10	3.76	0.59	2.35	1.92	21.25	IV
41-1	100—102	375	2.90	6.47	10.39	0.10	3.16	1.11	1.65	1.46	6.79	IV
42-2	129—131	386	2.50	5.54	9.69	0.08	3.88	0.40	2.07	1.97	4.94	IV
44-2	80—82	405	2.21	6.19	7.68	0.07	4.81	2.70	1.35	1.65	17.07	IV
45-2	110—113	415	2.07	6.53	9.47	0.05	4.56	3.81	2.90	1.15	9.31	IV
46-5	110—112	429	1.60	6.16	7.95	0.09	4.33	1.14	2.57	2.09	15.44	IV
48-2	91—94	443	1.15	8.47	8.03	0.08	4.98	1.28	3.24	0.96	5.89	IV
49-2	84—87	452	0.75	6.78	6.67	0.05	6.32	0.91	2.50	1.12	1.90	IV
51-2	10—13	471	1.80	5.93	8.19	0.08	3.63	1.12	2.03	1.68	6.54	IV
52-2	50—52	481	1.43	6.70	6.69	0.07	4.10	1.13	2.35	1.49	14.07	IV
52-4	126—128	485	0.80	7.62	6.81	0.09	5.07	0.61	3.66	0.29	6.62	V
53-2	137—140	491	0.74	7.96	6.21	0.10	4.65	0.98	3.95	0.19	7.07	V
54-5	115—117	505	0.81	7.98	7.28	0.08	5.73	1.34	2.48	0.30	2.90	V
55-2	129—131	510	1.08	4.35	10.42	0.12	3.07	2.22	1.29	0.93	31.21	V
56-2	50—52	519	1.04	2.89	4.29	0.06	1.89	0.67	1.02	0.59	9.88	VI
57-2	85—87	547	2.35	5.42	8.00	0.16	4.60	1.21	1.28	1.29	16.73	VI

TABLE II

DSDP Site 253 — Middle Eocene volcanogenic sequence — Trace element analyses (ppm)

Core section	Interval (cm)	Depth below sea floor (m)	Li	Sr	Ba	Cr	Mn	Co	Ni	Cu	Zn	Mo	Lithological units (McKelvey and Fleet, 1974)
17-2	60—62	154	50	39	178	73	10,832	22	90	98	129	10	I
18-1	60—62	162	46	335	125	83	2,249	39	75	54	162	8	I
19-2	9—11	172	13	200	82	88	1,551	40	78	69	150	5	I
20-2	90—92	182	28	426	41	44	1,167	51	62	60	74	4	I
21-2	58—60	192	57	6	22	10	232	15	46	21	15	7	I
21-4	112—114	195	72	48	22	33	570	49	49	81	113	2	II
23-2	86—88	211	25	53	21	108	1,046	50	133	90	54	4	II
26-2	23—25	239	57	23	21	89	384	64	115	85	180	4	II
28-2	18—19	252	27	39	21	75	3,927	42	84	143	56	4	II
30-1	120—122	271	9	456	10	73	1,012	39	64	29	97	4	III
31-2	118—120	281	34	146	73	93	666	60	81	136	93	8	III
36-1	93—95	327	65	158	1673	20	64	16	34	60	105	4	III or IV
38-1	101—103	347	3	82	104	53	815	36	64	122	123	8	III or IV
40-2	30—32	366	2	48	31	111	969	27	80	89	98	4	IV
41-1	100—102	375	4	142	41	26	955	65	65	85	217	5	IV
42-2	129—131	386	5	78	n.d.	29	775	53	56	126	115	6	IV
44-2	80—82	405	9	37	44	90	726	67	118	89	89	4	IV
45-2	110—113	415	4	51	21	61	487	56	82	134	101	5	IV
46-5	110—112	429	2	64	31	85	853	65	88	129	116	5	IV
48-2	91—94	443	6	33	<10	101	656	66	171	162	74	5	IV
49-2	84—87	452	16	56	n.d.	261	536	62	186	114	53	2	IV
51-2	10—13	471	5	270	61	145	764	49	95	171	80	5	IV
52-2	50—52	481	<2	449	51	110	654	55	87	100	96	5	IV
52-4	126—128	485	6	7	167	298	923	38	160	148	44	7	V
53-2	137—140	491	5	11	42	149	999	56	141	145	81	4	V
54-5	115—117	505	10	17	32	152	755	54	148	127	65	5	V
55-2	129—131	510	<2	299	36	75	1,238	20	68	91	61	5	V
56-2	50—52	519	9	23	41	37	618	41	45	64	55	<1	VI
57-2	85—87	547	4	30	43	64	1,637	49	79	114	111	3	VI

n.d. = not determined.

TABLE III

DSDP Site 253 — Middle Eocene volcanogenic sequence — Major element analyses on a calcium carbonate-free basis (%)

Core section	Interval (cm)	Depth below sea floor (m)	Ti	Al	Fe	Mn	Mg	Ca	Na	K	Lithological units (McKelvey and Fleet, 1974)
17-2	60—62	154	1.97	6.60	11.41	1.10	2.26	1.29	2.83	2.57	I
18-1	60—62	162	1.87	5.80	8.73	0.42	3.84	2.67	3.12	1.66	I
19-2	9—11	172	1.84	5.64	10.20	0.24	3.11	4.89	2.64	2.27	I
20-2	90—92	182	1.72	5.04	7.43	0.23	3.81	7.95	3.22	1.07	I
21-2	58—60	192	0.07	5.44	10.82	0.02	5.10	1.99	2.59	0.85	I
21-4	112—114	195	1.55	7.00	8.11	0.07	3.96	2.10	2.53	1.48	II
23-2	86—88	211	0.91	7.97	6.41	0.12	3.85	3.29	2.47	1.76	II
26-2	23—25	239	1.17	8.35	6.51	0.04	4.68	3.97	2.54	0.86	II
28-2	18—19	252	1.08	7.85	5.68	0.48	4.69	2.63	2.31	0.83	II
30-1	120—122	271	1.07	4.81	11.22	0.17	3.81	3.43	1.75	2.61	III
31-2	118—120	281	2.00	7.00	8.37	0.07	4.47	2.20	2.31	1.10	III
36-1	93—95	327	0.44	7.40	4.31	0.01	4.02	1.89	2.57	0.71	III or IV
38-1	101—103	347	2.61	6.54	9.52	0.09	3.80	2.07	2.74	1.13	III or IV
40-2	30—32	366	1.84	6.91	9.56	0.13	4.77	0.75	2.98	2.44	IV
41-1	100—102	375	3.11	6.94	11.15	0.11	3.39	1.19	1.77	1.68	IV
42-2	129—131	386	2.63	5.83	10.19	0.08	4.08	0.42	2.18	2.07	IV
44-2	80—82	405	2.66	7.46	9.26	0.08	5.80	3.26	1.63	1.99	IV
45-2	110—113	415	2.28	7.20	10.44	0.06	5.03	4.20	3.20	1.27	IV
46-5	110—112	429	1.89	7.28	9.40	0.11	5.12	1.35	3.04	2.47	IV
48-2	91—94	443	1.22	9.00	8.53	0.09	5.29	1.36	3.44	1.02	IV
49-2	84—87	452	0.76	6.91	6.80	0.05	6.44	0.93	2.55	1.14	IV
51-2	10—13	471	1.93	6.34	8.76	0.09	3.88	1.20	2.17	1.80	IV
52-2	50—52	481	1.66	7.80	7.79	0.08	4.77	1.32	2.73	1.73	IV
52-4	126—128	485	0.86	8.16	7.29	0.10	5.43	0.65	3.92	0.31	V
53-2	137—140	419	0.80	8.57	6.68	0.11	5.00	1.05	4.25	0.20	V
54-5	115—117	505	0.83	8.22	7.50	0.08	5.90	1.38	2.55	0.31	V
55-2	129—131	510	1.57	6.32	15.15	0.17	4.46	3.23	1.88	1.35	V
56-2	50—52	519	1.15	3.21	4.76	0.07	2.10	0.74	1.13	0.65	VI
57-2	85—87	547	2.82	6.51	9.61	0.19	5.52	1.45	1.54	1.55	VI

TABLE IV

DSDP Site 253 — Middle Eocene volcanogenic sequence — Trace element analyses on a calcium carbonate-free basis (ppm)

Core section	Interval (cm)	Depth below sea floor (m)	Li	Sr	Ba	Cr	Mn	Co	Ni	Cu	Zn	Mo	Lithological units (McKelvey and Fleet, 1974)
17-2	60—62	154	51	40	182	74	11,047	22	92	100	132	10	I
18-1	60—62	162	87	633	236	157	4,251	73	142	81	306	15	I
19-2	9—11	172	20	305	125	134	2,364	61	119	105	229	8	I
20-2	90—92	182	53	816	78	84	2,234	98	119	115	142	8	I
21-2	58—60	192	58	6	22	10	236	15	47	21	15	7	I
21-4	112—114	195	79	53	24	36	627	54	54	89	124	2	II
23-2	86—88	211	31	65	26	132	1,281	61	163	110	66	5	II
26-2	23—25	239	58	23	21	91	391	65	117	87	183	4	II
28-2	18—19	252	33	48	27	92	4,800	51	103	175	68	5	II
30-1	120—122	271	16	789	17	126	1,752	68	111	50	168	7	III
31-2	118—120	281	36	156	78	100	713	64	87	146	100	9	III
36-1	93—95	327	66	160	1694	20	65	16	34	61	106	4	III or IV
38-1	101—103	347	4	96	122	62	958	42	75	143	145	9	III or IV
40-2	30—32	366	3	61	39	141	1,230	34	102	113	124	5	IV
41-1	100—102	375	4	152	44	28	1,025	70	70	91	233	5	IV
42-2	129—131	386	5	82	n.d.	31	815	56	59	133	121	6	IV
44-2	80—82	405	11	45	53	109	875	81	142	107	107	5	IV
45-2	110—113	415	4	56	23	67	537	62	90	147	148	6	IV
46-5	110—112	429	2	76	37	101	1,009	77	104	153	137	6	IV
48-2	91—94	443	6	35	—	107	697	70	182	172	79	5	IV
49-2	84—87	452	16	57	n.d.	266	546	63	190	116	54	2	IV
51-2	10—13	471	5	289	65	155	817	52	102	183	86	5	IV
52-2	50—52	481	—	523	59	128	761	64	101	116	112	6	IV
52-4	126—128	485	6	7	179	319	988	41	171	158	47	7	V
53-2	137—140	491	5	12	45	160	1,075	60	152	156	87	4	V
54-5	115—117	505	10	17	33	157	778	56	152	130	67	5	V
55-2	129—131	510	—	434	52	109	1,800	29	99	132	89	7	V
56-2	50—52	519	10	26	45	41	686	45	50	71	61	—	VI
57-2	85—87	547	5	36	52	77	1,966	59	95	137	133	4	VI

n.d. = not determined.

TABLE V

DSDP Site 253 — Middle Eocene volcanogenic sequence — Some average trace element contents for the lithological units on a calcium carbonate-free basis (standard deviations for data shown in brackets)

Lithological unit	Li	Ba	Cr	Co	Ni	Cu	Zn	Mg
I	54	129	92	54	104	84	169	—
	(24)	(84)	(57)	(35)	(36)	(38)	(110)	
II	50	25	88	58	109	115	110	—
	(23)	(3)	(39)	(6)	(45)	(41)	(55)	
III	39	—	82	56	77	86	125	—
	(25)		(55)	(15)	(39)	(52)	(38)	
IV	<6*	<55*	109	61	111	134	122	—
	(4)	(30)	(67)	(14)	(43)	(29)	(47)	
V	<7*	77	186	47	144	144	73	—
	(3)	(68)	(92)	(14)	(31)	(15)	(20)	
VI	8	49	59	52	73	104	97	—
	—	—	—	—	—	—	—	
I, II and III	49	76**	88	54	99	95	137	3.97
	(22)	(75)	(47)	(24)	(39)	(41)	(78)	(0.76)
IV, V and VI	<6*	<61*	121	57	114	133	108	4.75
	(4)	(41)	(78)	(15)	(43)	(29)	(46)	(1.05)

* Average does not include sample with contents beneath detection limit.
** Excluding sample 36-1, 93—95.

TABLE VI

DSDP Site 253 — Middle Eocene volcanogenic sequence — Some average major element contents for each unit on a calcium carbonate-free basis (standard deviations for data shown in brackets)

Lithological unit	Ti	Al	Fe	Mg	Na	K
I	1.49	5.70	9.66	3.62	2.88	1.68
	(0.80)	(0.58)	(1.55)	(1.05)	(0.28)	(0.74)
II	1.18	7.79	6.68	4.30	2.46	1.23
	(0.27)	(0.57)	(1.02)	(0.45)	(0.11)	(0.46)
III	1.17	6.40	7.97	4.10	2.21	0.87
	(0.78)	(1.39)	(3.47)	(0.34)	(0.42)	(0.59)
IV	2.05	7.11	9.22	4.76	3.02	1.70
	(0.69)	(0.83)	(1.22)	(0.92)	(1.53)	(0.52)
V	1.02	7.82	9.16	5.20	3.15	0.54
	(0.37)	(1.01)	(4.01)	(0.61)	(1.12)	(0.54)
VI*	1.99	4.86	7.19	3.81	1.10	1.10
	—	—	—	—	—	—

* Average of only two samples which have dissimilar compositions.

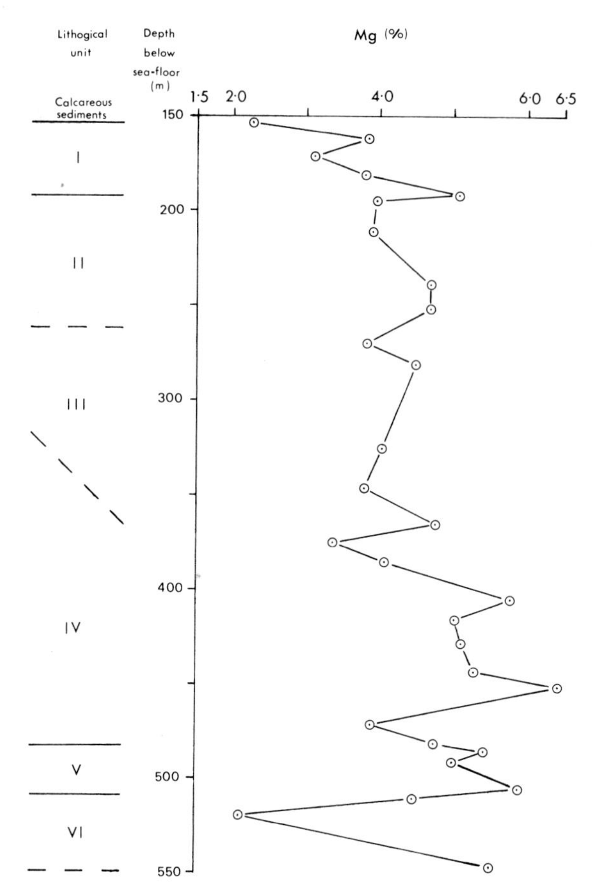

Fig.5. The variation of Mg, on a $CaCO_3$-free basis, with depth in the volcanogenic sequence at Site 253.

I and III). This apparent systematic variation is also indicated by the relatively Mg-rich smectite, saponite, being present in the lower three units and not in the upper three (McKelvey and Fleet, 1974). Similarly, Ni, Cr, and Cu are enriched in the lower relative to the upper part of the sequence (Table V; Figs.6 and 7) whilst Li and Zn are relatively depleted (Table V; Fig.8). All these element variations support the suggestion of Fleet that the upper and lower parts of the sequence were derived from parent magmas of different composition (McKelvey and Fleet, 1974). The thin quartz-normative basalt flow in Unit II at 219 m and, in contrast, the Mg-rich highly olivine-normative picritic basalt at 588 m at the base of the drilled sequence (Kempe, 1974; Frey and Sung, 1974) also indicates a change in magma composition during formation of the sequence. By inference the relatively higher Mn, Ni, and Cr, and lower Li and Zn contents of the lower three units specifically implies the derivation of these units from basalt of more basic composition than that which gave rise to the upper three units.

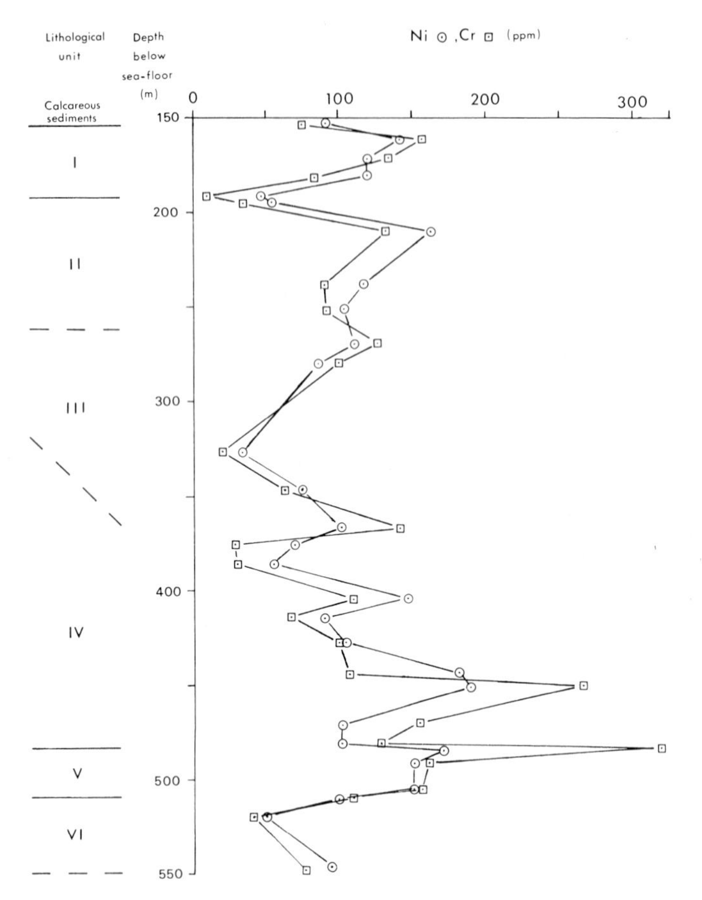

Fig.6. The variations of Ni and Cr, on $CaCO_3$-free bases, with depth in the volcanogenic sequence at Site 253.

(2) The considerable and varied post-eruptive alteration of the units and the limited number and irregular intervals of the samples implied that apparent differences in the contents of the individual units must be treated with caution; for instance the two samples from Unit VI are very different in composition (Tables I—IV). Even over short distances in a volcanogenic sequence the effect of alteration on the chemistry of pyroclastic material may vary significantly (e.g. Hay and Iijima, 1968). The differing effects of all these factors is reflected by the high standard deviations of the data for each unit (Table V and VI). Only the average Ti and Fe contents appear to vary from unit to unit. Units VI and IV contain about 2% Ti, Units V, II, and III about 1%, and Unit I approximately 1.5% (Table VI). Similarly, the average Fe contents of the units varies: Units I, IV and V have over 9%, Unit III has about 8%, and Units II and IV have 7%. Whilst the varying Ti and Fe contents of the units may result from differentiation, they could as well result from the differing hydrothermal and/or diagenetic alteration of each unit, though each unit may not have undergone similar alteration throughout.

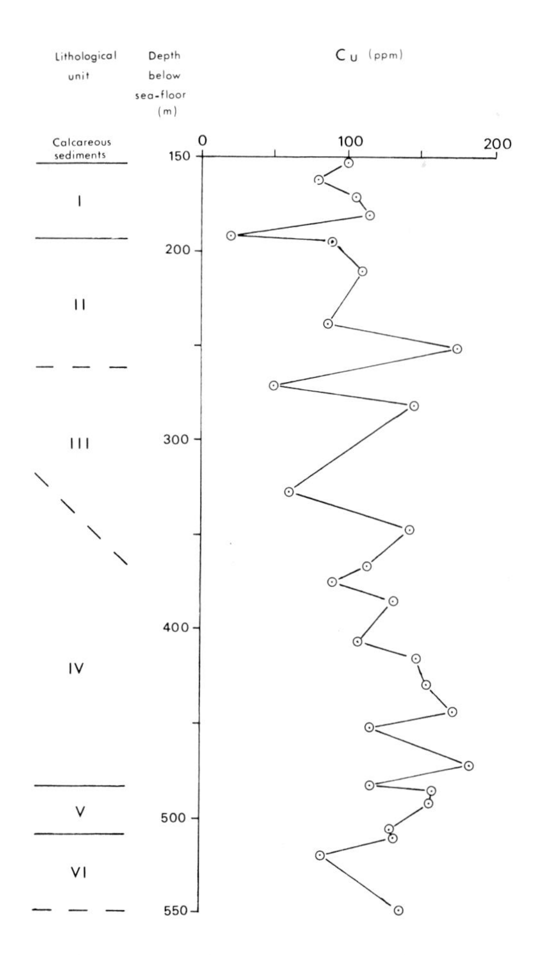

Fig.7. The variation of Cu, on a $CaCO_3$-free basis, with depth in the volcanogenic sequence at Site 253.

(3) The unsystematic variations with depth exhibited by Al, Na, K, Ca, Ba and Sr (Tables V and VI) are not surprising. Christensen et al. (1973), for example, have pointed out that the low temperature alteration of glass may result in the gain *or* loss of these elements, and therefore the high degree of alteration the sequence has undergone is liable to have resulted in such unsystematic variations. Deposition of salts from porewater before and during drying will also affect the contents of some elements, especially sodium. The Sr abundances (Table II) are possibly governed by the carbonate contents of the samples, whilst the Co contents vary only slightly (Table IV).

(4) The Mn contents of the samples decrease rapidly with depth (Tables III and IV, Fig.9), so that below 190 m nearly all the samples contain less than 2000 ppm Mn. The selective dissolution studies indicate that Mn is variously fractionated in the selected samples analysed (Table VII). In samples 17-2, 60—62 and 28—2, 18—19, which contain significantly more Mn

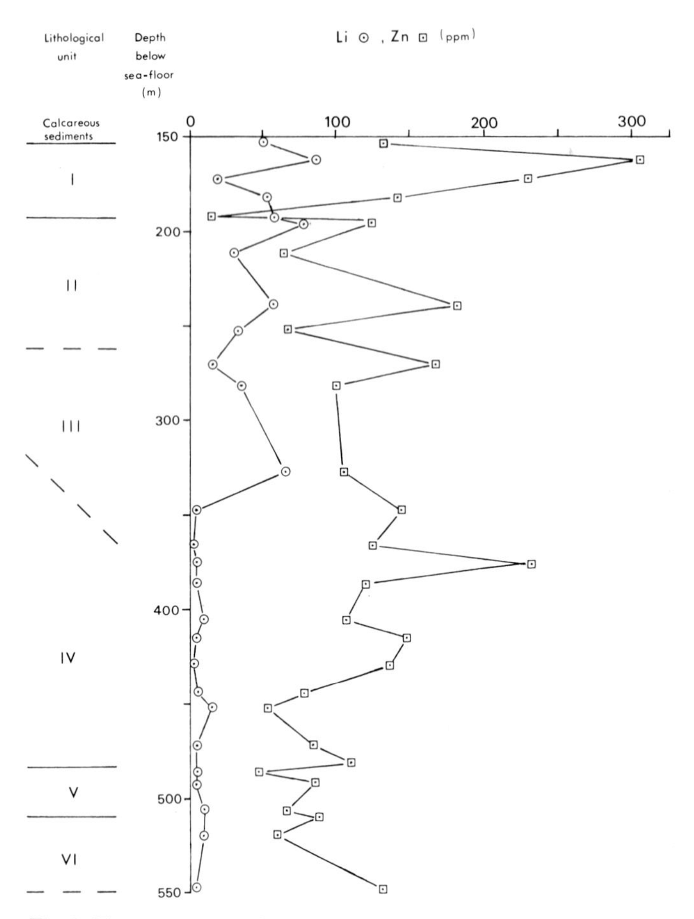

Fig.8. The variations of Li and Zn, on $CaCO_3$-free bases, with depth in the volcanogenic sequence at Site 253.

than most of the others, about 75% of the Mn present is soluble in the mixed acid-reducing agent, and is therefore contained in Fe—Mn oxyhydroxides. Only minor quantities of the Mn in samples 38-1, 101—103 and 42-2, 129—131, however, is present in such oxyhydroxides. 72% of the Mn in sample 52-4, 126—128 is either in readily exchangeable sites or is present in carbonates (i.e. is soluble in the acetic acid), though the latter alternative seems unlikely as the $CaCO_3$ content of the sample is only about 7.0% (Table I). Little of the Mn in any of the samples would seem to be in resistant minerals as 80—90% of the Mn is soluble in the hydrochloric acid. Ni and Cu, which are usually concentrated in Fe—Mn oxyhydroxides (e.g. Chester, 1965), are fractionated differently from each other and from Mn (Table VII). Little of the Ni in each sample is soluble in the mixed acid-reducing agent, but most dissolves in the HCl. The Ni in the analysed samples therefore appears to be neither in Fe—Mn oxyhydroxides nor in chemically resistant minerals. Cu, in contrast, is varyingly fractionated. In the three analysed samples (38-1,

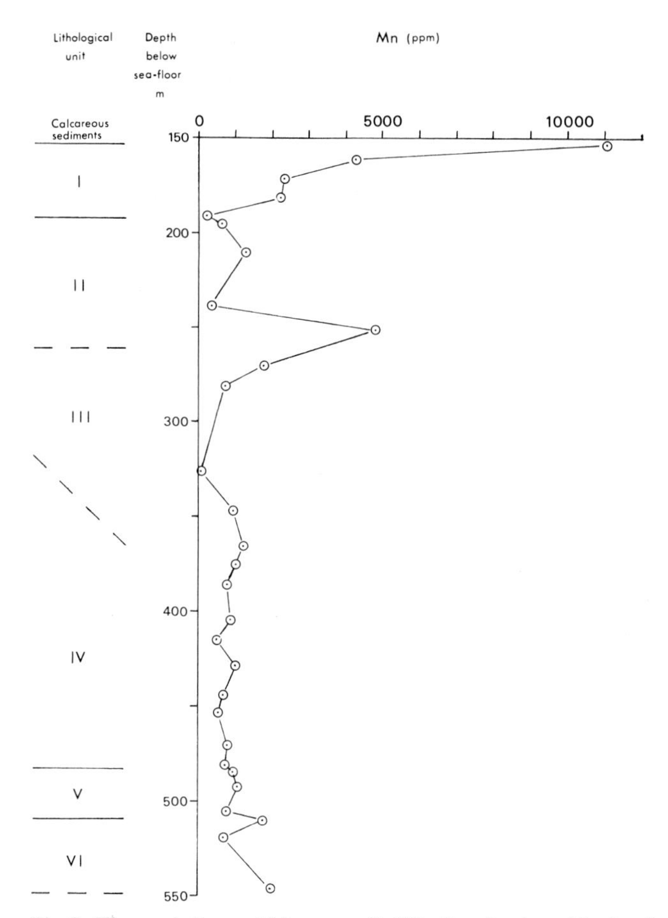

Fig.9. The variation of Mn, on a $CaCO_3$-free basis, with depth in the volcanogenic sequence at Site 253.

101—103; 42-2, 129—131; and 52-4, 126—128) which contain the least Mn about half the Cu is soluble in acetic acid, and most of it is associated with Fe—Mn phases. The other two samples, in which substantial amounts of Fe—Mn oxyhydroxides are present, contain little Cu in readily exchangeable sites or carbonates, and 54% and 28% of their Cu is associated with their Fe—Mn minerals.

The variation in Mn content with depth strongly suggests two phases of upwards migration through the sequence during diagenesis. The relatively high value at 252 m would appear to be related to the weathered interval at the top of unit III at 262.2 m and so represents the older period of migration. We suggest that closer sampling about the latter interval would reveal perhaps an even higher concentration, developed there during the period of non-deposition and submarine weathering. The fact that the high peak recorded at 252 m is just within the base of Unit II indicates subsequent renewed upwards migration of the Mn concentration. This later migration could be

TABLE VII

Results of selective fraction dissolution of volcanogenic samples from DSDP Site 253 (amount of element dissolved both in ppm and as a percentage of total content)

Dissolution agent / Sample	Elements									
	Mn				Ni			Cu		
	25% (v/v) CH_3COOH	mixed acid-reducing agent	hot 50% (v/v) HCl	HF	mixed acid-reducing agent	hot 50% (v/v) HCl	HF	25% (v/v) CH_3COOH	mixed acid-reducing agent	HF
17-2, 60—62	466 4%	7948 73%	9655 89%	10832 —	<15 <17%	70 78%	90 —	<15 <15%	53 54%	98 —
28-2 18—19	n.d. —	3086 79%	3386 86%	3927 —	17 20%	62 73%	85 —	12 8%	40 28%	143 —
38-1, 101—103	262 32%	291 36%	671 82%	815 —	<15 <22%	60 90%	67 —	50 41%	114 93%	122 —
42-1, 129—131	196 25%	248 32%	681 88%	775 —	<15 <30%	51 100%	50 —	76 60%	121 96%	126 —
52-4, 127—128	665 72%	729 79%	821 89%	923 —	<15 <9%	147 89%	165 —	64 43%	126 85%	148 —

n.d. = not determined.

approximately contemporaneous with the upwards migration of Mn evident in Unit I.

Mn is known to migrate from zones of reducing conditions at depth and to be deposited under oxidising conditions near surface. Lyn and Bonatti (1965) found that this occurs over a few metres distance in hemipelagic sediments, and that although sulphides may exist at depth they are not invariably present. They therefore concluded that some Mn would migrate even under "slightly reducing conditions". This may have been the case in the Site 253 sequence, or at least in Unit I, as no significant sulphide concentrations are present. An additional factor enhancing mobilisation may have been that Mn was present in exchange sites, at least in the material of Unit I. Comparison of the Site 253 sequence with that of Site 254 near the southern end of the Ninetyeast Ridge leads to this suggestion. The latter sequence is entirely epiclastic and is thought to consist of sub-aerially eroded material (Davies et al., 1974, Ch.6). Although it contains sulphides no Mn mobilisation appears to have occurred in it (Fleet, unpublished data), possibly because all the readily mobilised Mn had been lost during erosion.

Comparison with related basalts. The sequence is enriched in the light relative to the heavy rare earth elements (REE) but the REE contents vary unsystematically with depth, probably because of high-temperate subaqueous alteration and the presence of biogenous calcite (Fleet et al., 1976). The very nature of the formation of the hyaloclastics is likely to have led to far greater high-temperature alteration, than occurs when submarine lavas are erupted and rapidly chilled. The REE contents of the volcanogenic sediments from Sites 254 and 258 (Naturaliste Plateau), and the basalts from the Ninetyeast Ridge are similarly fractionated (Thompson et al., 1974; Frey and Sung, 1974; Fleet et al., 1976), except for some samples from the basal basalt at Site 253, which are depleted in the light relative to the heavy REE (Frey, personal communication, 1976). The REE data from Site 253, therefore, indicates that the sequence resulted from the kind of volcanism responsible for the whole of the Ninetyeast Ridge, and, probably, the Naturaliste Plateau (Fleet et al., 1976). This volcanism results in rocks which are relatively rich in incompatible elements, and is therefore distinct from that which produces the great majority of mid-ocean ridge basalts. Along the Ninetyeast Ridge and Naturaliste Plateau it may have resulted from two fixed hot spots (Luyendyk et al., in prep.), but no wholly satisfactory explanation of the geochemical characteristics of this kind of volcanism have been advanced, even where far more data are available (e.g. O'Nions et al., 1976).

Whilst all the igneous rocks from the Ninetyeast Ridge have relatively high contents of incompatible elements, only the Site 253 basalts are distinguished from ocean ridge tholeiites solely by this characteristic (Frey and Sung, 1974). They, themselves, are divisible into quartz and highly olivine-normative tholeiites (Kempe, 1974; Frey and Sung, 1974). As shown above the geochemical variation exhibited by the Site 253 hyaloclastic and epiclastic sediments may be correlated with the compositions of these two types of basalt. The sequence, therefore, seems to be the product of volcanism contiguous with that which produced the Site 253 basalts. Hence the magmas from which the Site 253 volcanic suite formed, whilst being generally similar to those extruded elsewhere on the Ninetyeast Ridge, are distinct in detail.

CONCLUSIONS

(1) The Middle Eocene hyaloclastic sediments at Site 253 are largely the result of violent basaltic hydroexplosions, occurring in water depths of less than 150 m.

(2) The presence of subaerially erupted basaltic ash-fall tuffs interbedded within the younger part of the hyaloclastic sequence marks the emergence of some Ninetyeast Ridge volcanic vents above sea level.

(3) The volcanogenic sequence is divisible into two geochemically distinct parts, the less basic upper part composed of Units I to III and the more basic lower part composed of Units IV to VI.

(4) The respective chemistries of these upper and lower parts correlate with that of the thin quartz-normative basalt within Unit II; and the basal Mg-rich olivine-normative basalt beneath Unit VI.

(5) The volcanogenic sequence is enriched in light relative to heavy rare earth elements, as are very nearly all igneous rocks from the Ninetyeast Ridge. This shows the volcanism that produced the Site 253 sequence is like that responsible for the whole of the Ninetyeast Ridge.

(6) The volcanogenic sequence shows marked upwards migration of Mn during diagenesis.

ACKNOWLEDGEMENTS

The authors thank Drs Paul Henderson and David Kempe for critically reading an early draft of the geochemical section of the paper. The geochemistry reported here was carried out while one of the authors (A.J.F.) was in receipt of a N.E.R.C. Joint Research Studentship at Chelsea College, University of London, and the British Museum (Natural History); this grant is gratefully acknowledged.

REFERENCES

Bonatti, E., 1970. Deep sea volcanism. Naturwissenschaften, 57 (8): 379—384.

Buchanan, D.J. and Dullforce, T.A., 1973. Mechanism for vapour explosions. Nature, 245: 32—34.

Carlisle, D., 1963. Pillow breccias and their aquagene tuffs, Quadra Island, British Columbia. J. Geol., 71: p.45.

Chester, R., 1965. Elemental geochemistry of marine sediments. In: J.P. Riley and G. Skirrow (Editors), Chemical Oceanography, 2. Academic Press, New York, N.Y., pp.23—80.

Chester, R. and Hughes, M.J., 1967. A chemical technique for the separation of ferromanganese minerals, carbonate minerals and adsorbed trace elements from pelagic sediments. Chem. Geol., 2: 249—262.

Christensen, N.I., Frey, F.A., MacDougall, D., Melson, W.G., Peterson, M.N.A., Thompson, G. and Watkins, N.D., 1973. Deep Sea Drilling Project; properties of igneous and metamorphic rocks of the oceanic crust: EOS Trans. AGU, 55: 972—982.
Colgate, S.A. and Sigurgeirsson, T., 1973. Dynamic mixing of water and lava. Nature, 224: 552—555.
Cronan, D.S. and Garrett, D.E., 1973. Distribution of elements in metalliferous Pacific sediments collected during the Deep Sea Drilling Project. Nature, Phys. Sci., 242: 88—89.
Cucuzza Silvestri, S., 1963. Proposal for a genetic classification of hyaloclastites. Bull. Volcanol., 25: 315—321.
Davies, T.A., Luyendyk, B.P., Rodolfo, K.S., Kempe, D.R.C., McKelvey, B.C., Leidy, R.D., Horvath, G.J., Hyndman, R.D., Thierstein, H.R., Herb, R.C., Boltovshoy, E. and Doyle, P., 1974. Initial Reports of the Deep Sea Drilling Project, 26. U.S. Govt. Printing Office, Washington, D.C., 1129 pp.
Fleet, A.J., Henderson, P. and Kempe, D.R.C., 1976. Rare earth element and related chemistry of some drilled southern Indian Ocean basalts and volcanogenic sediments. J. Geophys. Res., 81: 4257—4268.
Frey, F.A. and Sung, C.M., 1974. Geochemical results for basalts from Sites 253 and 254. In: T.A. Davies, B.P. Luyendyk et al., Initial Reports of the Deep Sea Drilling Project, 26. U.S. Govt. Printing Office, Washington, D.C., pp.567—572.
Hay, R.L. and Iijima A., 1968. Nature and origin of palagonite tuffs of the Honolulu Group on Oahu, Hawaii. In: R.R. Coats, R.L. Hays and C.A. Anderson (Editors), Studies in Volcanology. Geol. Soc. Am., Mem., 116: 331—376.
Heiken, G., 1972. Morphology and petrology of volcanic ashes. Geol. Soc. Am. Bull., 83: 1961—1988.
Kempe, D.R.C., 1974. The petrology of the basalts, Leg 26. In: T.A. Davies, B.P. Luyendyk et al., Initial Reports of the Deep Sea Drilling Project, 26. U.S. Govt. Printing Office, Washington, D.C., pp.465—503.
Luyendyk, B.P., Rennick, W. and Sclater, J.G., in preparation. Tectonic origin of aseismic ridges in the eastern Indian Ocean.
Lyn, D.C. and Bonatti, E., 1965. Mobility of manganese in diagenesis of deep-sea sediments. Mar. Geol., 3: 457—574.
McBirney, A.R., 1963. Factors governing the nature of submarine volcanism. Bull. Volcanol., 26: 455—469.
McKelvey, B.C. and Fleet, A.J., 1974. Eocene basaltic pyroclastics at Site 253, Ninetyeast Ridge. In: T.A. Davies, B.P. Luyendyk et al., Initial Reports of the Deep Sea Drilling Project, 26. U.S. Govt. Printing Office, Washington, D.C., pp.553—565.
Milliman, J.D., 1974. Marine Carbonates. Springer, New York, N.Y., 375 pp.
O'Nions, R.K., Pankhurst, R.J. and Grönvold, K., 1976. Nature and development of basalt magma sources beneath Iceland and Reykjanes Ridge. J. Petrol., 17: 315—338.
Peckover, R.S., Buchanan, D.J. and Ashby, D.E.T.F., 1973. Fuel—coolant interactions in submarine volcanism. Nature, 245: 307—308.
Platania, G., 1891. Geological notes of Acireale. In: H.J. Johnston-Lavis (Editor), The south Italian Volcanoes. Furcheim, Noakes.
Platania, G., 1902—03. Acicastello — Ricerche geologiche e vulcanologiche. Mem. Class. Sci. Acc. Zelanti, Ger. 3, Vol. 2, Acireale.
Rittman, A., 1962. Volcanoes and Their Activity. Wiley, New York, N.Y.
Sartorius v. Walterhausen, W., 1846. Über die submarinen vulkanischen Ausbruche in der Tertier Formation des Val di Noto. Gottinger Studien 1945, Gottingen.
Sartorius v. Walterhausen, W., 1853. Über die vulkanischen Gesteine in Sicilien und Island und ihre submarine Umbildung. Gottingen.
Scheidegger, K.F., 1973. Volcanic ash layers in deep sea sediments and their petrological significance. Earth Planet. Sci. Lett., 17: 397—407.
Tazieff, H., 1972. About deep-sea volcanism. Geol. Rundsch., 61 (2): 470—480.
Thompson, G., Bryan, W.B., Frey, F.A. and Sung, C.M., 1974. Petrology and geochemistry of basalts and related rocks from Sites 214, 215, 216, DSDP Leg 22, Indian Ocean. In: C.C. von der Borch, J.G. Sclater et al., Initial Reports of the Deep Sea Drilling Project, 22. U.S. Govt. Printing Office, Washington, D.C., pp.459—468.
Verhoogen, J., 1951. Mechanics of ash formations. Am. J. Sci., 249: 729—739.

Marine Geology, 26 (1978) 99—117

GEOCHEMISTRY OF VOLCANICS FROM THE NINETYEAST RIDGE AND ITS VICINITY IN THE INDIAN OCEAN

V. VISWANATHA REDDY[1], K.V. SUBBARAO[1], G.R. REDDY[2], J. MATSUDA[3], and R. HEKINIAN[4]

[1] *Indian Institute of Technology, Powai, Bombay 400 076 (India)*
[2] *Bhabha Atomic Research Centre, Trombay, Bombay 400 085 (India)*
[3] *Geophysical Institute, Tokyo University, Tokyo (Japan)*
[4] *Centre océanologique de Bretagne, 29273 Brest (France)*

(Received March 28, 1977)

ABSTRACT

Viswanatha Reddy, V., Subbarao, K.V., Reddy, G.R., Matsuda, J. and Hekinian, R., 1978. Geochemistry of volcanics from the Ninetyeast Ridge and its vicinity in the Indian Ocean. Mar. Geol., 26: 99—117

The basalts and oceanic andesites from the aseismic Ninetyeast Ridge display trachytic, vesicular and amygdaloidal textures suggesting a subaerial volcanic environment. The normative composition of the Ninetyeast Ridge ranges from olivine picrite- to nepheline-normative alkaline basalt, suggesting a wide range of differentiation. This is further supported by the fractionation—differentiation trends displayed by transition metal trace elements (Ni, Cr, V and Cu). The Ninetyeast Ridge rocks are enriched in rare earth (RE) and large ion lithophile (LIL) elements and Sr isotopes (0.7043—0.7049), similar to alkali basalts and tholeiites from seamounts and islands, but different from LIL-element-depleted tholeiitic volcanic rocks of the recent seismic mid-Indian oceanic ridge. The constancy of $^{87}Sr/^{86}Sr$ ratios for basalts and andesites is compatible with a model involving fractional crystallization of mafic magma. The variation of $(La/SM)_{ef}$ ratios between 0.97 and 2.79 may possibly be explained in terms of a primordial hot mantle and/or chemically contrasting heterogeneous mantle source layers relatively undepleted in LIL elements at different periods in the geologic past. In general, the Sr isotopic data for rocks from different tectonic environments are consistent with a "zoning—depletion model" with systematically arranged alternate alkali-poor and alkali-rich layers in the mantle beneath the Indian Ocean.

INTRODUCTION

During Legs 22 and 26 of the Deep Sea Drilling Project (DSDP) in the eastern Indian Ocean floor, several holes reaching the basement had been drilled on the Ninetyeast Ridge. This ridge is a north—south linear feature, extending from 5° N to about 30° S and rising 1500—2000 m above the surrounding sea floor. Because of its structure and its aseismic nature, this ridge is of particular interest since similar features are present elsewhere in other oceans and their origin is still controversial. Wilson (1965) and Morgan

(1971, 1972) have suggested that aseismic ridges throughout the world are chains of elevated volcanoes which are the surface expression resulting from plates moving away from fixed hot spots in the mantle. However, the asymmetric nature between the eastern and the western slope—basin ocean floor intersection of the Ninetyeast Ridge suggests that its eastern part is associated with a transform fault type of structure (Bowin, 1973; McKenzie and Sclater, 1971).

Previous studies carried out on the rocks collected from the DSDP in the eastern Indian Ocean dealt mainly on the petrology and geochemistry with particular emphasis on the relationship between various tectonic provinces and the compositional variations of their exposed volcanics. It was mainly emphasized that the crustal composition of the Ninetyeast Ridge was different from that found on the recent mid-Indian spreading ridges and instead is comparable to that found on some islands of the Indian Ocean such as New Amsterdam and St. Paul (Hekinian, 1974a). This is further substantiated by the fact that lignite and shallow-water organisms were discovered in certain drill sites (e.g. 214 and 216; Von der Borch et al., 1974). The presence of highly vesicular and amygdaloidal textural features found in some levels of the drillhole also suggest a subaerial volcanic environment. Based on the trace-element analyses of rocks from Sites 214 and 216, Thompson et al., (1974) suggested that the fractionated lava of the Ninetyeast Ridge may represent the surface expression of a former mantle plume with consequent uplift of the lithosphere as it moved over the hot spot. Although DSDP Sites 214, 215 and 216 of Leg 22 (Fig.1) were previously studied for major and transitional metal element and REE distribution (Hekinian, 1974a; Bougault, 1974; Thompson et al., 1974; Kempe, 1974), new data at different levels of the drillholes emphasizing mostly the distribution of large ion lithophile elements (LIL) and of Sr isotopic ratios are presented in this paper. Such additional information permits a better evaluation of magmatic evolution and the relationship between the various lithological units. This in turn facilitates understanding the units and allows inferences to be made on the different mantle source material.

METHODS OF STUDY

Rare-earth (La, Ce, Sm, Eu, Tb and Yb), Cs and Rb concentrations were determined using neutron-activation analysis techniques involving radiochemical group separations (Reddy et al., 1976) at the Bhabha Atomic Research Centre, Trombay, Bombay. Both Na I (Tl) as well as Ge (Li) detectors coupled with multichannel analyzers were used. The analytical uncertainties range from 2 to 7%. Geological rock standards were repeatedly analyzed to determine the precision and accuracy of our methods of analysis. The results are presented in Table I.

Ni, Cr, V and Cu determinations were made by S. Ramanan using the optical emission spectrographic method, at the Bhabha Atomic Research Centre, Trombay, Bombay and the experimental details are presented by

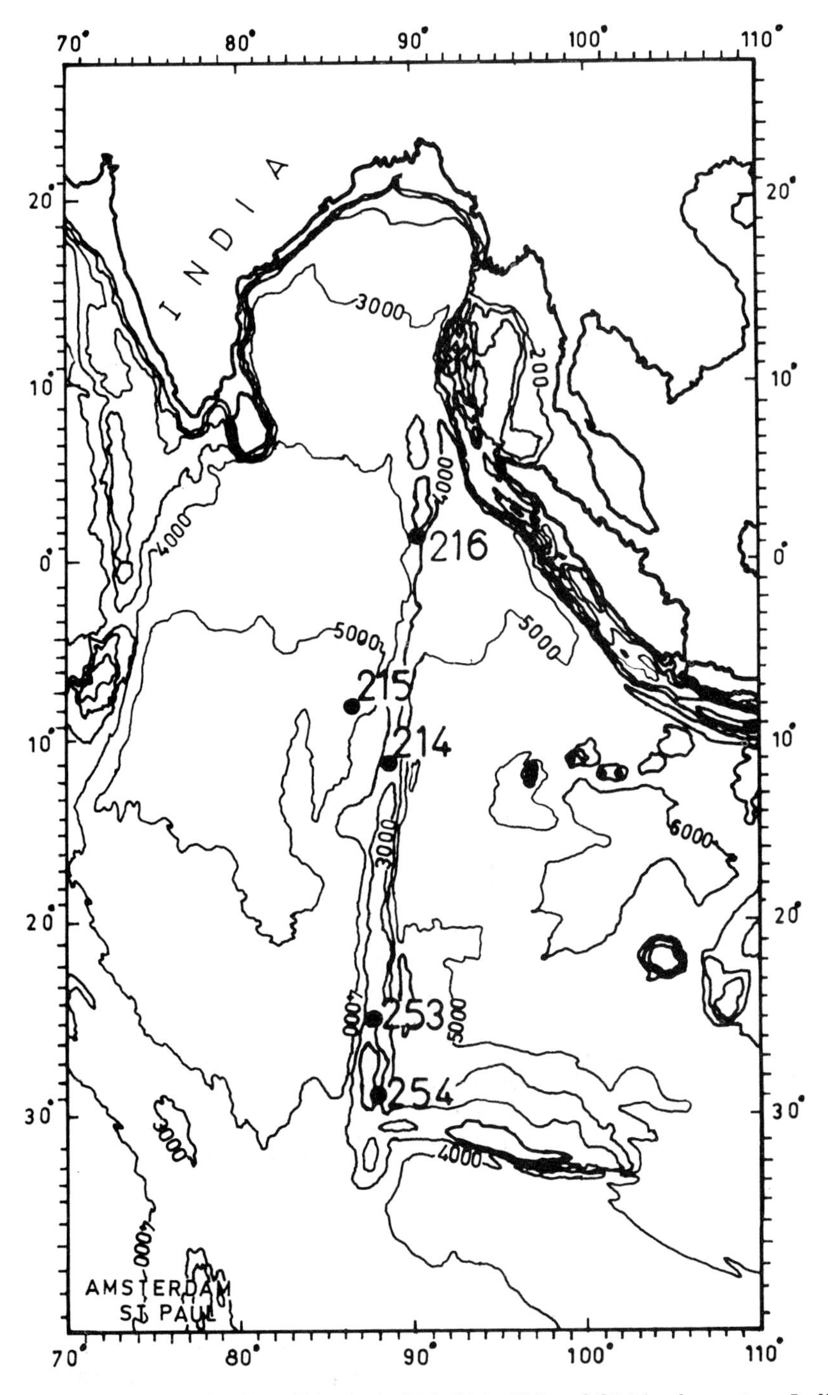

Fig.1. Location of DSDP Sites 214, 215, 216, 253 and 254 in the eastern Indian Ocean (Legs 22 and 26).

TABLE I

Analytical data for USGS rock standards*

	G-2	GSP-1	AGV-1	BCR-1
Cs	1.3 (1.4)	1.1 (1.0)	1.25 (1.4)	0.9 (0.95)
Rb	170 (168)	254 (254)	65.9 (67.0)	46.5 (46.6)
La	87.9 (96.0)	179 (191)	37.2 (35.0)	29.9 (26.0)
Ce	160 (150)	413 (394)	63.0 (63.0)	53.4 (53.9)
Sm	7.76 (7.3)	33.4 (27.1)	7.39 (5.9)	6.89 (6.6)
Eu	1.6 (1.5)	2.7 (2.4)	1.9 (1.7)	2.0 (1.94)
Tb	0.32 (0.54)	2.8 (1.3)	0.89 (0.70)	1.08 (1.0)
Yb	—	—	2.0 (1.7)	—

* Values in brackets are from Flanagan (1973).
All values in ppm.

Ramanan (1976). The analytical uncertainties range from 5 to 10%. USGS rock standards were constantly used to determine the precision and accuracy of our results.

The Sr isotopic ratios were measured on a 25-cm radius single-focussing mass spectrometer using a single tantalam filament mode of ionisation, with a digital output (analytical precision is ± 0.0002 at 2σ), at Tokyo University, Japan. The $^{87}SR/^{86}SR$ ratio of E&A $SrCO_3$ is 0.7080.

We have obtained slightly higher $^{87}Sr/^{86}Sr$ ratios for volcanic rocks from the Ninetyeast Ridge which may probably be due to the low to medium degree of alteration in this case. These ratios may be slightly reduced if these samples are subjected to HCl washing before the analysis.

SAMPLE DESCRIPTIONS AND MAJOR-ELEMENT STUDY

Detailed descriptions of the various drillholes from Leg 22 permitted us to recognize various lithological units within the individual sites (Fig. 2) (Hekinian, 1974a,b). The rock types encountered along the Ninetyeast Ridge and on Site 215 located nearby in the Central Indian Basin consist of pyroxene-enriched basalts, also called ferrobasalts by Frey et al. (1976), amygdalar and vesicular basalt with plagioclase phenocrysts (Site 253; Kempe, 1974) and also intermediate differentiated rocks (Hekinian, 1974b).

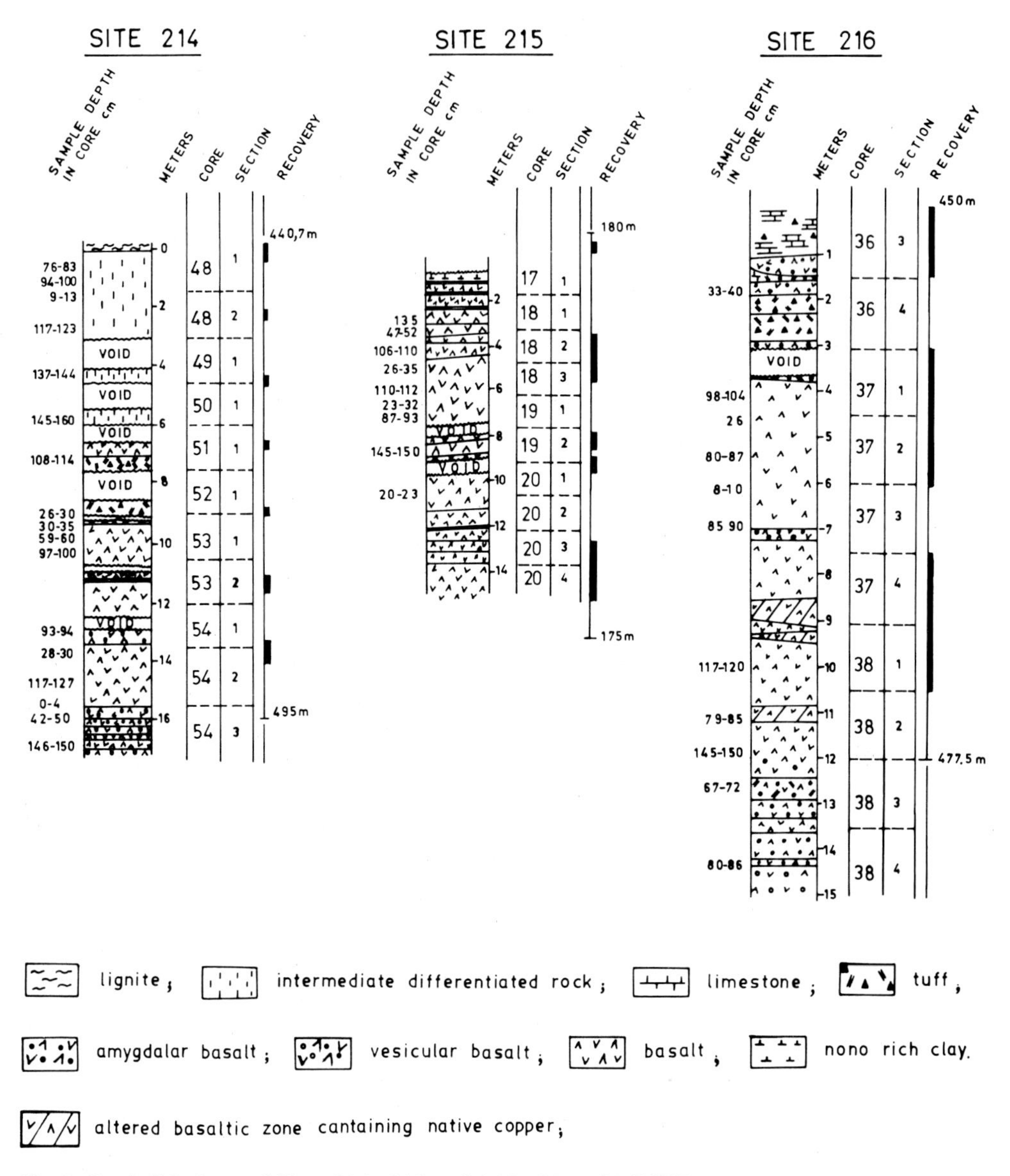

Fig.2. Rock lithology of Sites 214, 215 and 216 of Leg 22,DSDP.

These intermediate differentiated rocks from Site 214 are characterized by the presence of higher TiO_2, Fe_2O_3 + FeO, FeO/MgO, and lower SiO_2, than continental or island-arc calc-alkaline suites (Table II). Therefore these have been termed "oceanic andesites". The picritic basalt (Site 253;

TABLE II

Partial major-element data of volcanic rocks from Sites 214, 216 and 215

	Site 214 (6 samples)	Site 216 (7 samples)	Site 215 (6 basalts)	MIOR (33 basalts)	Oceanic andesites, Site 214 (8 samples)	Calc-alkaline rocks (6 samples)
SiO_2:						
Range	42.7—47.7	45.7—50.4	47.5—50.2	45.9—51.4	54.59—58.0	50.6—66.8
Average	44.9	48.4	49.9	49.3	56.13	59.2
TiO_2:						
Range	2.02—2.50	2.50—2.76	1.63—1.67	0.07—2.19	1.39—1.56	0.23—1.13
Average	2.17	2.68	1.66	1.14	1.44	0.77
K_2O:						
Range	0.18—0.81	0.44—1.29	0.80—0.96	0.09—0.64	1.27—1.83	1.07—2.67
Average	0.34	0.75	0.87	0.25	1.53	1.74
Na_2O:						
Range	1.85—2.95	2.30—2.63	2.86—3.7	2.15—3.51	3.70—4.21	2.89—4.97
Average	2.47	2.49	3.07	2.67	3.95	3.89
K_2O/Na_2O:						
Range	0.07—0.23	0.15—0.34	0.26—0.32	0.03—0.20	0.33—0.44	0.33—0.73
Average	0.12	0.23	0.28	0.09	0.39	0.45
Fe_2O_3+FeO:						
Range	13.2—15.7	13.1—15.0	8.22—8.90	4.66—14.6	6.52—10.4	2.27—8.74
Average	14.5	14.0	8.51	8.76	9.55	6.11
FeO/MgO*						
Range	2.16—2.45	1.88—2.70	1.17—1.44	0.46—1.75	3.73—6.02	0.93—1.87
Average	2.27	2.19	1.31	1.12	4.23	1.33

Sources: *Sites 214, 215 and 216*: Hekinian (1974a); MIOR: Engel et al. (1965), Hekinian (1968), Balashov et al. (1970), Engel and Fisher (1975); *Calc-alkaline rocks*: Jakes and White (1972).

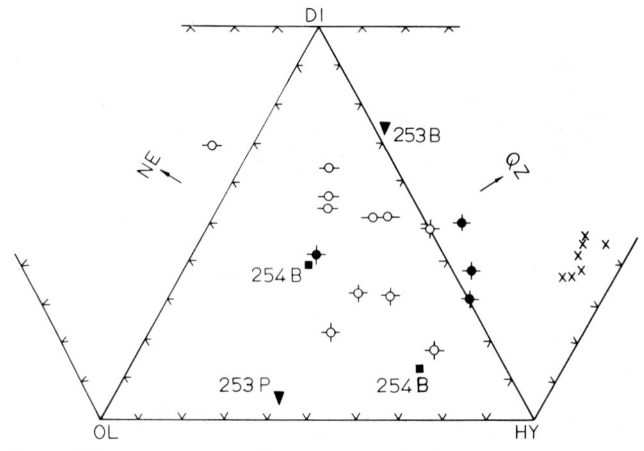

Fig.3. Normative plots of volcanic rocks from Sites 214, 215, 216, 253 and 254. Norms are calculated with a fixed Fe_2O_3 at 1.5%. Leg 22 analyses are from Hekinian (1974a) (⌖ basalt and × oceanic andesite, Site 214; -o- basalt, Site 215; ◆ basalt, Site 216), and Leg 26 analyses are from Kempe (1974) (B-basalt and P-picrite).

Kempe, 1974) found on the Ninetyeast Ridge differs from that collected from recent spreading ridges by its higher total-iron content (Hekinian and Thompson, 1976).

The various rock types found on the Ninetyeast Ridge display wide compositional range from quartz-normative to nepheline-normative basalts in the normative tetrahedron of Yoder and Tilley (1962) (Fig.3), suggesting differentiation. In general, the major-element chemistry of aseismic Ninetyeast Ridge rocks resembles that of alkali basalts from islands and seamounts, but differs significantly from tholeiitic volcanics of the seismic mid-Indian Oceanic ridge (MIOR).

TRACE-ELEMENT AND STRONTIUM ISOTOPE GEOCHEMISTRY

Rare-earth elements (REE)

Rare-earth element data (La, Ce, Sm, Eu, Tb, Yb) for eight basalts and two andesites from Sites 214, 215, 216 are presented in Table III. We have also used published data of La and Sm (Thompson et al., 1974) for different samples of basalts from these sites which we have not analyzed, to obtain a nearly complete genetic history of these sites. Although the chondrite-normalized REE patterns of these rocks (Fig.4) are uniformly enriched in light rare-earth elements (LREE), the absolute REE enrichment varies by a factor of three. This is further supported by the large variation in $(La/Sm)_{e.f}$ ratio (0.97—2.8). In general, the LREE enriched patterns of the Ninetyeast Ridge basalts strongly contrast with the LREE depleted patterns of most of the mid-oceanic ridge basalts (Fig.4). The REE patterns for Site 214 basalts and andesites display certain interesting features:

(1) Two geochemical groups are apparent within the basalts (Fig.4)—$(La/Sm)_{e.f}$ ratios vary from 0.97 to 2.8 (Table III).

(2) There is a close match of the highly LREE enriched pyroxene basalt with that of the LREE enriched andesites (Fig.4). This is further corroborated by higher $(La/Sm)_{e.f}$ ratios for this basalt (2.8) and andesite (1.94) (Table III).

(3) The REE patterns of basalts and andesites are similar to tholeiites and alkali basalts from St. Paul, Hawaii and Réunion islands (Gast, 1968; Schilling and Winchester, 1969; Kay and Gast, 1973; Zielinski, 1975; Frey et al., 1976; Table III and Fig.5). Therefore, we may have a situation on the Ninetyeast Ridge like that in the above-mentioned islands, e.g., similar mode and type of volcanism and differentiation, as well as similar mantle source areas and magma depth.

K, Rb, and Cs variations

The basalts and andesites from the Ninetyeast Ridge are distinctly enriched in the alkali elements similar to alkali-island basalts but strongly contrast

TABLE III

Analytical data

	Site 214								Site 215			Site 216			Island basalts			
	basalts					oceanic andesites			basalts			basalts			tholeiites		alkali basalts	
	53–1 (26–30)	53–1 (32–35)	54–3 (0–4)	54–2 (28–30)	average of three basalts *A	48–1 (76–83)	49–2 (0–7)	average of four andesites *B	18–1 (68–73) (glass)	19–1 (87–93)	average of five basalts *C	37–2 (80–87)	38–2 (79–85)	average of two basalts *D	St. Paul (I)	Hawaii (II)	Hawaii (III)	MIOR (IV)
K	1743	2400	2158	—	—	12948	12948	—	6900	6640	16	6059	6889	—				1700
Rb	2.64	3.55	3.83	—	—	35.1	48.7	—	26.8	19.5	—	19.7	16.4	—				2.5
Cs	0.24	0.63	0.31	—	—	0.55	0.79	—	0.42	0.16	—	0.26	0.13	—				0.12
La	10.7	31.1	7.20	9.2	8.4	38.7	36.8	32.7	17.8	17.5	16.2	16.6	14.8	13.2	19.6	7.2	18	4.4
Ce	22.5	81.8	12.5	—	—	67.2	64.2	—	61.1	31.6	36.0	30.8	30.2	—	—	25.8	50	13.8
Sm	5.01	6.10	2.72	4.6	5.1	11.1	10.4	11.3	5.15	5.11	4.5	5.45	5.72	5.0	7.26	4.6	5.6	2.66
Eu	1.66	1.77	0.81	—	1.77	3.01	2.94	3.17	1.79	1.70	1.5	1.60	1.62	1.45	2.3	1.6	1.9	1.19
Tb	0.76	1.15	0.55	—	—	2.0	1.43	—	0.69	1.19	0.8	0.91	1.13	—	—	0.82	0.82	0.78
Yb	3.66	—	1.76	—	3.0	5.15	4.88	4.39	—	2.65	2.5	3.62	3.85	3.6	3.6	1.7	1.6	2.44
Ni		33			50	15		5		104	100		20					107
Cr		36			38	10		5		240	250		25					319
V		192			525	22		39		246	245		249					243
Cu		150			—	5		—		54	65		48					84
K/Cs	7263	3908	6961	—	—	23542	16390	—	16586	41500		23304	52992					20882
K/Rb	660	670	561	—	—	369	266	—	256	341		308	420					831
$^{87}Sr/^{86}Sr$	0.7046	0.07048	—	0.7043	—	0.7049	0.7049	—	0.7045			—	—					0.7034
$(La/Sm)_{e.f}$	1.11	2.79	1.46	1.1	0.97	1.94	1.94	1.59	1.89	1.89		1.67	1.42	1.45	1.48	0.86	1.76	1.06
$(La/Yb)_{e.f}$	1.77	—	2.48	—	1.79	4.55	4.57	4.71	—	4.0		2.78	2.33	2.35	3.3	2.57	6.81	1.18

* Frey et al., 1976 (A: 53–1(26–30), 53–1(97–100) and 54–2(117–127); B: 48–2(9–13), 48–2(111–123), 49–1(137–144) and 50–1(145–150); C: 18–2(106–110), 18–3(110–112), 19–1(23–32) and 20–2(20–23); D: 38–2(80–87) and 38–c/c).

(I) Frey et al.(1976); (II) Average of 5 basalts (Schilling and Winchester, 1969); (III) Average of 8 basalts (Schilling and Winchester, 1969; Kay and Gast, 1973); (IV) Average of 34 basalts (Engel and Fisher, 1969; Engel et al., 1974; Engel and Fisher, 1975 and our unpublished data).

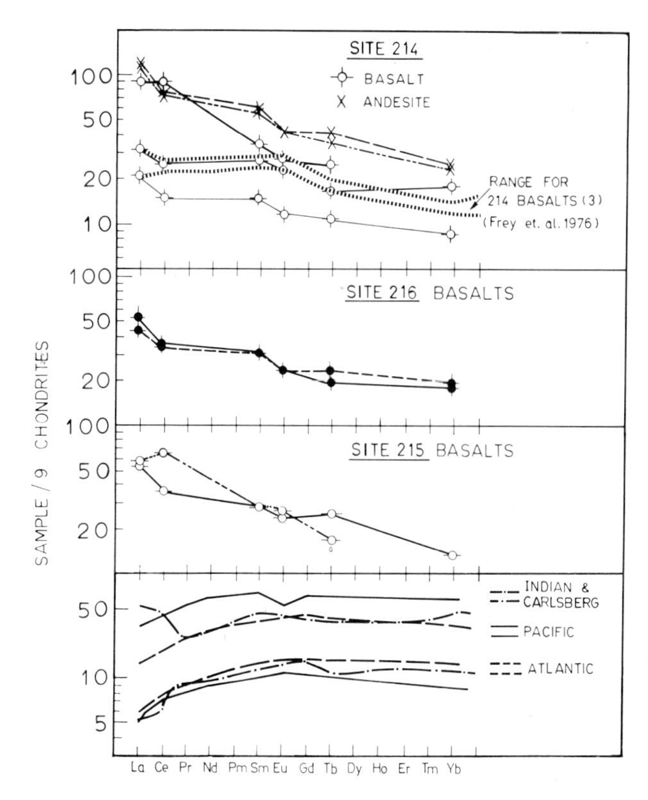

Fig.4. REE distribution patterns for 3 basalts (-o-) and 2 andesites (x) from Site 214; 2 basalts (●) from 216 and 2 basalts (-o-) from 215. The bottom diagram shows fields for mid-oceanic ridges (Source: MIOR & CR: unpublished data (Viswanatha Reddy) and Schilling (1971); MAR: Kay, Hubbard and Gast (1970); Schilling (1971); EPR (East Pacific, Juan de Fuca, and Chile Rises): Kay et al. (1970) and Frey and Haskin (1964).

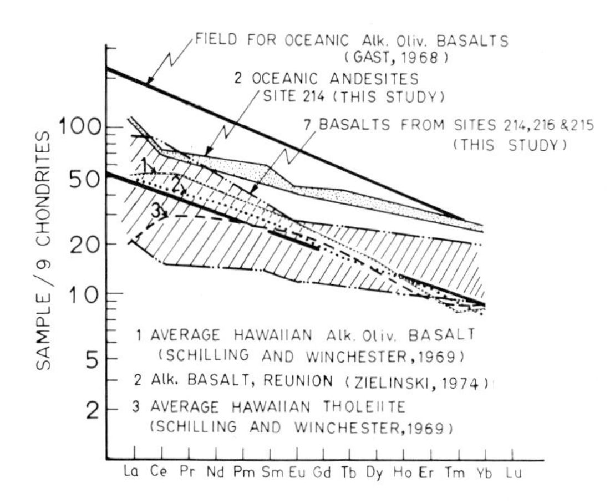

Fig.5. Diagram showing similarities between the REE patterns of Ninetyeast Ridge and island tholeiitic and alkali basalts.

with the depleted mid-ocean ridge basalts (Table III; Fig.6). In general, most of the alkali elements are sensitive to seawater alteration; therefore it is necessary to have relatively fresh samples in order to understand the real nature of the source rocks and differentiation trends.

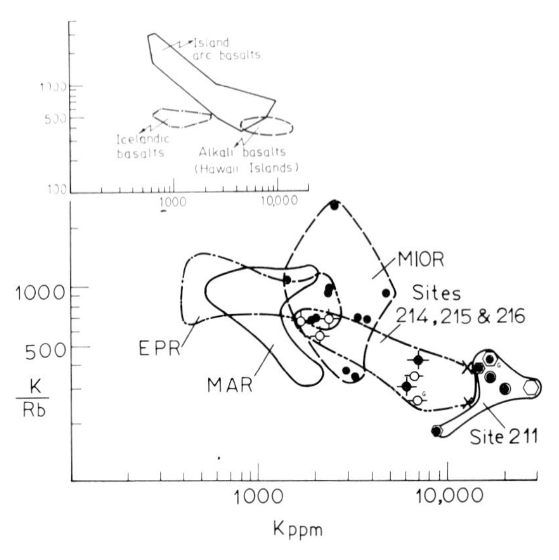

Fig.6. Variation of K/Rb vs. K for rocks from MIOR (Subbarao et al., 1976; Erlank et al., 1974), MAR and EPR (Hart, 1971) and Sites 214 (-◇-), 215 (-○-) and 216 (◆). Fields for Icelandic basalts, alkali-island basalts from Hawaii and island-arc basalts are from Hart (1976).

As most of the crystalline samples under study are altered at least to a certain degree, we have firstly used the REE concentrations which are believed to be insensitive to seawater alteration (Philpotts et al., 1969; Kempe and Schilling, 1974) to derive the real chemical nature of the source regions. All these rocks are uniformly enriched in LREE and alkali elements similar to alkali-island basalts. Therefore we assume that the samples under study are not too much altered to be excluded for geochemical investigations.

In order to speculate on the nature of the source areas, we have used a K/Rb—K plot, as Cs is relatively more affected by seawater alteration. The K/Rb—K variation of samples from the aseismic Ninetyeast Ridge as well as basalts from recent seismic mid-ocean ridge environments from the Pacific, Atlantic and Indian Oceans are shown in Fig.6. The Ninetyeast Ridge rocks, while following the general trend of decreasing K/Rb with increasing K, do in fact display a distinct identity from the seismic mid-ocean ridge basalts.

The basaltic glass from Site 215 has the lowest K/Rb ratio of 256, in contrast to higher ratios of 660 and 670 for the crystalline basalts from this series. In view of the fact that the seawater alteration lowers the K/Rb ratio, we consider the higher K/Rb to be significant, probably representing the composition of the source areas.

The following points merit consideration:

(1) The existence of two geochemical groupings of the Ninetyeast Ridge samples of Sites 214 and 216 and Site 215, with K/Rb ratios around 250 and 700 (Fig.6).

(2) The fact that basalts having higher K/Rb ratios around 700 partly overlap the fields of EPR and MIOR, while basalts and andesites having lower K/Rb ratios around 300—400 overlap the field of alkali-island basalts (Fig.6).

Thus the K/Rb—K plots of the Ninetyeast Ridge rocks show characteristics similar to mid-oceanic ridge basalts, island-arc basalts and alkali-island basalts. In other words, we have a situation with a heterogeneous mantle source(s) which was later subjected to mixing processes in order to produce intermediate basalts (olivine normative in the sense of Yoder and Tilley, 1962, their fig. 3). The negative trend of the Ninetyeast Ridge rocks may either represent a differentiation process by crystal fractionation or a heterogeneity in the mantle chemistry. The lower K/Rb ratio of the glassy basalt from Site 215 would definitely reflect the relatively undepleted nature of the mantle source. It seems likely that the relatively lower and higher K/Rb ratios around 250 and 700 may represent chemical variations in the mantle source layers which probably resulted due to different degrees of depletions in LIL elements. The higher $^{87}Sr/^{86}Sr$ ratios of 0.7043—0.7049 for these rocks are also generally consistent with a model involving the presence of a relatively undepleted mantle source. These ratios appear to be similar to those reported for other tectonically related areas such as islands, seamounts, abyssal hills but different from the seismic mid-oceanic ridges (Subbarao, 1972; Subbarao and Hekinian, 1978).

Although K/Rb—K plot of the Ninetyeast Ridge rocks does not offer definite clues to the genesis of these rocks, these data at least emphasize some unique similarities of the aseismic ridge rocks to rocks from other tectonic environments such as islands and seamounts, implying similar mantle sources and volcanic and differentiation processes. These characteristics in turn throw light on mantle differentiation.

Ni, Cr, V and Cu variations

We report here concentrations of Ni, Cr, V and Cu for four samples from Sites 214, 215 and 216 (Table III). We have also included average analyses of Frey et al. (1976) of volcanic rocks from these sites for comparison. These elements are considered to be useful indicators of crystal fractionation.

The aseismic Ninetyeast Ridge rocks are depleted in Ni, Cr, V and Cu and are distinctly different from the seismic mid-oceanic ridges in having higher concentrations of these elements. The trace-element study for Site 214 rocks provides excellent geochemical evidence for fractional crystallization (Fig.7; Table III). The V concentration decreases with increasing fractionation. The strong enrichment of V in magnetite crystallizing from a silicate liquid is well described (Duncan and Taylor, 1968; Taylor et al., 1969; Ewart et al., 1973,

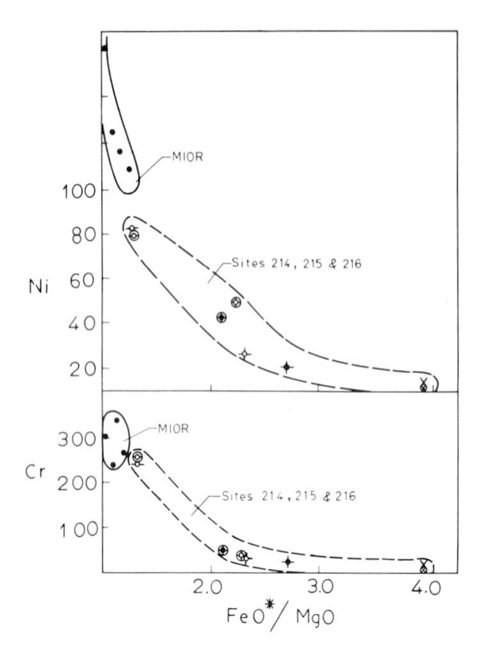

Fig.7. FeO*/MgO vs Ni and Cr variation diagram for volcanic rocks from MIOR (our unpublished data) and Ninetyeast Ridge (This study, and Engel et al., 1965). Symbols encircled are the averages from Frey et al. (1976).

Miyashiro and Shido, 1975). At Site 214 the magnetite-bearing ferrobasalt (FeO*/MgO = 2) has a V concentration of 192 ppm (Table III). After the magnetite fractionation the residual liquids would have been left with very low concentrations of V. This is evident from the low V concentration of 22 ppm in the overlying, differentiated oceanic andesite. These oceanic andesites appear to have been formed after the separation of magnetite and clinopyroxene from the ferrobasalts. This fact is also represented in Fig.7, which shows the fractionation of Cr and Ni with respect to FeO*/MgO. Furthermore, the andesite is also characterized by low concentrations of Cr, Ni, Cu and V than the underlying ferrobasalt.

Site 215 basalts are distinctly enriched in Ni, Cr, Cu and V, similar to mid-Indian oceanic ridge basalts (Fig.7, Table III). Although alkali-island basalts have lower concentrations of these elements, the island tholeiites have concentrations comparable to those of Site 215 basalts (Thompson et al., 1974).

The transition metal trace-element studies indicate the role of fractionation — differentiation in the formation of basalts and oceanic andesites from the Ninetyeast Ridge. Furthermore the concentration of these trace elements also shows a similarity to that of alkali basalt and tholeiite suites from islands.

---

*FeO is total iron as FeO.

$^{87}Sr/^{86}Sr$ variations

We also report here $^{87}Sr/^{86}Sr$ ratios for six samples from Sites 214 and 215 which show significant variation (Table III). Two basalts from Site 214 show a small range of variation, from 0.7046 to 0.7048, while the third sample yielded a lower ratio of 0.7043. On the other hand, the two andesites have higher $^{87}Sr/^{86}Sr$ ratios of 0.7049. Thus the nearly similar higher $^{87}Sr/^{86}Sr$ ratios for andesites and two basalts probably suggest a genetic relationship between these two rock types and is compatible with a model involving either fractional crystallization of a basaltic magma (relatively undepleted in LIL elements) or variable degrees of partial melting of a single mantle source. This relationship is also consistent with the major and minor element chemistry (Figs.3 and 7) particularly Ni, Cu, Cr, and V. However the significantly lower $^{87}Sr/^{86}Sr$ ratio of 0.7043 poses a problem. It is significant that the higher $^{87}Sr/^{86}Sr$ ratios (0.7043—0.7049) are definitely different from the recent seismically active mid-oceanic ridge rocks. (0.7021—0.7035; Peterman and Hedge, 1971) but similar to oceanic island basalts (0.7035—0.7055; Peterman and Hedge, 1971). Although these large Sr isotopic differences between the basalts (0.7043—0.7048) within a single site may possibly reflect the presence of two mantle sources, it is difficult to accept this interpretation in view of limited analytical data. The fact that $(La/Sm)_{e.f}$ varies widely between 0.97 and 2.79, with most of the ratios lying around 1.60, is significant, and probably consistent with a model involving the presence of hybrid or transitional mantle source(s).

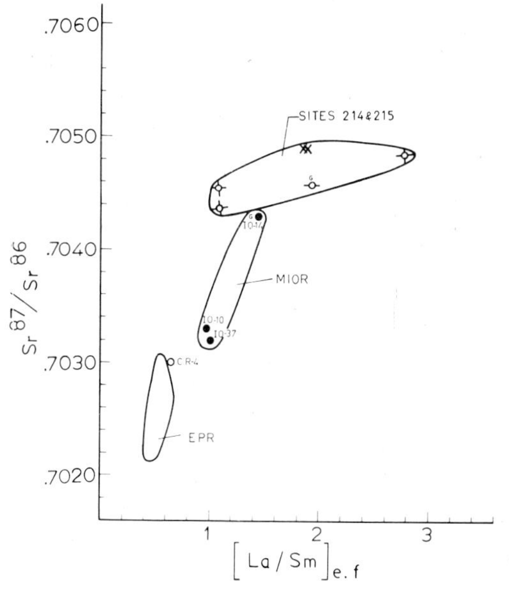

Fig.8. $^{87}Sr/^{86}Sr$ vs.$(La/Sm)_{e.f}$ variation diagram. Sites 214 and 215: This study; MIOR and Carlsberg Ridge (CR): Viswanatha Reddy (unpublished data); EPR: Kay et al. (1970); Hedge and Peterman (1970); Peterman and Hedge (1971).

The $^{87}Sr/^{86}Sr$ ratio of 0.7045 for the glassy basalt from Site 215 lies within the $^{87}Sr/^{86}Sr$ range of site 214 volcanic rocks (0.7043—0.7049). This observation is in accordance with higher REE and alkali-element concentrations and $(La/Sm)_{e.f}$ ratios (Table III; Fig.4).

The relation of $^{87}Sr/^{86}Sr$ ratios to $(La/Sm)_{e.f}$ in aseismic Ninetyeast Ridge rocks and seismic spreading mid-oceanic ridges is shown in Fig.8. A positive correlation is apparent. In view of the fact that the $(La/Sm)_{e.f}$ ratio is insensitive to seawater alteration and little affected by shallow-depth fractional crystallization during magma ascent (Schilling, 1973), we consider this ratio as a good indicator of fractionation of the light rare earths and also adequate to speculate on mantle source compositions. The $^{87}Sr/^{86}Sr$ ratios offer clues on the time(s) of depletion, while variations in $(La/Sm)_{e.f}$ reflect various degrees of depletion in the mantle sources. It is significant that the mantle beneath EPR is relatively more depleted than the mantle beneath MIOR as shown by $^{87}Sr/^{86}Sr$ and $(La/Sm)_{e.f}$ variation trends (Fig.8). Furthermore, the depletion of the source of MIOR basalts may have occurred more recently and/or to a lesser degree than the basalts from other oceans. It seems likely that the mantle source of the Ninetyeast Ridge rocks are relatively undepleted in LIL elements compared with the recent spreading ridges like the EPR and the MIOR.

The $(La/Sm)_{e.f}$ ratios display large variations ranging from about 1 to 3, and strongly contrast with the seismic spreading mid-oceanic ridges (<1; Schilling, 1973). In terms of the classification proposed by Schilling (1973) based on $(La/Sm)_{e.f}$ ratios, the Ninetyeast Ridge rocks appear to have been derived from a primordial hot mantle plume as well as hybrid mantle sources, similar to some of the islands, seamounts and MIOR ($(La/Sm)_{e.f}$ 1.03). In other words, the Ninetyeast Ridge rocks are characterized mostly by the presence of alkali and tholeiitic-alkali transitional types of volcanism. These results imply the presence of chemically contrasting upper mantle source layers beneath the Ninetyeast Ridge.

Comparison of Sr isotope data

Based on the major-element composition, Hekinian (1974b) has shown that the volcanics from aseismic ridges including the Ninetyeast Ridge, the Cocos, the Iceland—Faeroe, and the Walvis ridges, display a trend of differentiation which has progressed further than is commonly encountered on mid-oceanic ridge rocks. Further, these are characterized by alkali volcanism similar to islands which are near to or associated with the ridges.

As a corollary to this study, in this paper we have used LIL elements and Sr isotopes for rocks from different tectonic environments such as seismic and aseismic ridges, and islands of the Indian Ocean to see whether or not any systematic variation exists. Fig.9 shows the distribution of available $^{87}Sr/^{86}Sr$ ratios for rocks from the Indian Ocean. It is apparent

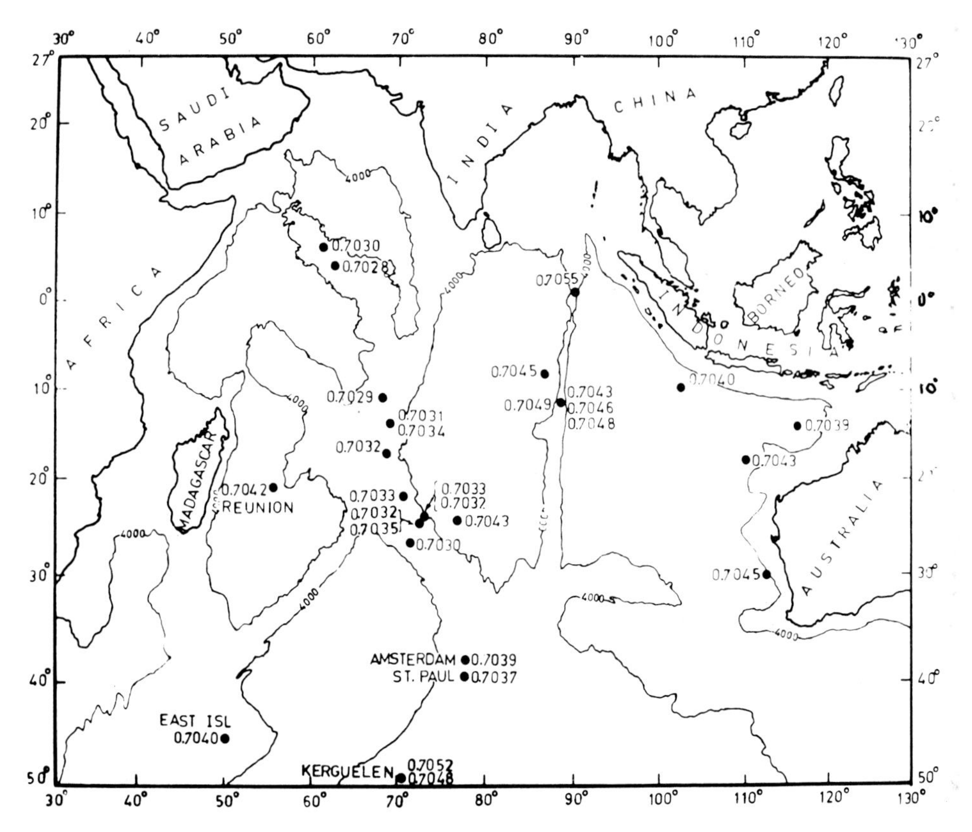

Fig.9. Map of the Indian Ocean showing the distribution of $^{87}Sr/^{86}Sr$ ratios from different tectonic environments (sources: DSDP Leg 22, Sites 211 and 214, 215 and 216: This study and our unpublished data; Leg 27, Sites 259, 260 and 261: Whitford (1975); MIOR: Subbarao and Hedge (1973); Engel and Fisher (1975); CR: Subbarao et al. (1976); Reunion, East, Amsterdam, St. Paul and Kerguelen islands: Hedge et al. (1973).

that the seismic mid-Indian oceanic ridge basalts consistently contain less radiogenic Sr (0.7029—0.7035; one value 0.7043) than the aseismic Ninetyeast Ridge (0.7043—0.7049). It is also evident that the aseismic Ninetyeast Ridge (0.7043—0.7049), and oceanic islands (0.7039—0.7052), have significantly higher $^{87}Sr/^{86}Sr$ ratios, suggesting similar mantle sources and volcanic processes. We interpret these systematic variations to the presence of alkali-poor and alkali-rich zones in the mantle giving rise to a chemically and isotopically zoned mantle. In other words, a zone having, say, about 0.7040 $^{87}Sr/^{86}Sr$ ratio may perhaps be feeding an island, a seamount, an abyssal hill, or an aseismic ridge.

An attempt has been made to compute mantle compositions of the Ninetyeast Ridge volcanic rocks using REE distributions of the rocks under study and REE partition coefficients for olivine, orthopyroxene, clinopyroxine, garnet and plagioclase. REE partitioning during partial melting of multiphase systems for three mantles (i.e. lherzolite, garnet—peridotite and plagioclase—peridotite) have been computed using the theoretical batch—equilibrium partial melting equations, to reconstruct the primary mantle REE patterns. This work is now in progress and the results will be published elsewhere.

SUMMARY

(1) The Ninetyeast Ridge (DSDP, Legs 22 and 26, Sites 214, 215, 216, 253, 254) consists of pyroxene basalts, amygdalar—vesicular basalts, picrites and intermediate differentiated rocks "oceanic andesites". These rocks are characterized by the presence of trachytic texture, in contrast to the quench textures exhibited by the mid-oceanic ridge rocks. The vesicular and amygdaloidal features indicate a subaerial environment.

(2) The normative composition of the aseismic Ninetyeast Ridge ranges from olivine picrite to nepheline-normative alkali basalt in the normative tetrahedron of Yoder and Tilley (1962), suggesting a wide range of differentiation. This is further supported by the fractionation trend displayed by Ni, Cr, V and Cu.

(3) The major-element chemistry resembles mostly alkali-island basalts suggesting alkali volcanism, in contrast to the tholeiitic volcanism of the seismic mid-Indian oceanic ridge.

(4) The oceanic andesites from Site 214 are characterized by the presence of higher TiO_2, Fe_2O_3 + FeO, FeO*/MgO, and lower SiO_2, compared with continental or island arc calc-alkaline rocks.

(5) The major-element chemistry of Site 215 glassy and crystalline basalts also resembles that of alkali basalts, lying between the ranges of basalts from Sites 214 and 216.

(6) The chondrite-normalized REE patterns of basalts and andesites are uniformly enriched in LREE, but the absolute REE enrichment varies by a factor of three. These are similar to alkali-island basalts and island tholeiites. The variation in REE, K/Rb (300 to 400, 700) and $(La/Sm)_{e.f}$ ratios (0.97 to 1.89, 2.79) indicate the possible presence of two groups of volcanic rocks within the Ninetyeast Ridge.

(7) The $^{87}Sr/^{86}Sr$ ratios of two basalts range from 0.7046 to 0.7048, while the third has a much lower value of 0.7043. Two andesites have higher values of 0.7049. The nearly similar higher $^{87}Sr/^{86}Sr$ ratios for two basalts and andesites is interpreted in terms of a model involving fractional crystallization of a basaltic magma. The variation of $(La/Sm)_{e.f}$ ratios between 0.97 to 2.79 may suggest the presence of chemically contrasting heterogeneous mantle source layers relatively undepleted in LIL elements, in contrast to recent spreading mid-oceanic ridges. The higher $(La/Sm)_{e.f}$ ratios also indicate the possible presence of primordial hot mantle and/or hybrid or transitional mantle source(s).

(8) The $^{87}Sr/^{86}Sr$ ratios of the aseismic Ninetyeast Ridge rocks resemble ratios from Indian Ocean island rocks, but different from those of seismic

mid-Indian oceanic ridges which have lower ratios. This is interpreted in terms of a "zoning—depletion model", with systematically arranged alternating alkali-poor and alkali-rich layers in the mantle giving rise to a chemically and isotopically zoned mantle beneath the Indian Ocean.

ACKNOWLEDGEMENTS

We are extremely grateful to M. Sankar Das, Bhabha Atomic Research Centre, Trombay, Bombay and M. Ozima, Tokyo University, Tokyo, for providing neutron activation and mass spectrometric laboratory facilities respectively as well as for reviewing the manuscript.

We are also indebted to the DSDP organizers for providing Leg 22 rock samples. Thanks are also due to S. Ramanan, Indian Institute of Technology, Bombay, for providing optical spectrographic analyses of Ni, Cr, Cu and V for eight samples.

REFERENCES

Balashov, Yu.A., Dimitriev, L.V. and Sharaskin, A.Ya., 1970. Distribution of the rare earths and Yttrium in the bedrock of the ocean floor. Geokhimiya, 6: 647—660 (in Russian, translated in Geochem. Int., 7: 456—468).

Bougoult, H., 1974. Distribution of first series transition metals in rocks recovered during DSDP Leg 22 in the north-eastern Indian Ocean. In: Initial Reports of the Deep Sea Drilling Project, 22. U.S. Govt. Printing Office, Washington, D.C., pp. 449—457.

Bowin, C.V., 1973. Origin of the Ninetyeast Ridge from studies near the Equator. J. Geophys. Res., 78: 6029—6043.

Duncan, A.R. and Taylor, S.R., 1968. Trace element analyses of magnetites from andesitic and dacitic lavas from Bay of Plenty, New Zealand. Contrib. Mineral. Petrol., 20: 30—33.

Engel, C.G. and Fisher, R.L., 1975. Granitic to ultramafic rock complexes of the Indian Ocean ridge system, Western Indian Ocean. Geol. Soc. Am. Bull., 86: 1553—1578.

Engel, C.G., Fisher, R.L. and Engel, A.E.J., 1965. Igneous rocks of the Indian Ocean floor. Science, 150: 605—610.

Engel, C.G., Bingham, E. and Fisher, R.L., 1974. Trace element compositions of Leg 24 basalts and one diabase. In: Initial Reports of the Deep Sea Drilling Project, 24. U.S. Govt. Printing Office, Washington, D.C., pp. 781—786.

Ewart, A.W., Bryan, W.B. and Gill, J.B., 1973. Mineralogy and geochemistry of the younger volcanic islands of Tonga, S.W. Pacific. J. Petrol., 14: 425—465.

Erlank, A.J., Reid, D.L. and Vallier, T.L., 1974. Petrology of Leg 25 basalts. In: E.S.W. Simpson et al., Initial Reports of the Deep Sea Drilling Project, 25. U.S. Govt. Printing Office, Washington, D.C.

Flanagan, F.J., 1973. 1972 values for international geochemical references samples. Geochim. Cosmochim. Acta, 37: 1189—1200.

Frey, F.A. and Haskin, L.A., 1964. Rare earths in oceanic basalts. J. Geophys. Res., 69: p. 775.

Frey, F.A., Dickey Jr., J.S., Thompson, G. and Bryan, W.D., 1976. Eastern Indian Ocean DSDP sites: correlations between petrography, geochemistry and tectonic setting, Geol. Soc. Am. Mem., in press.

Gast, P.W., 1968. Trace element fractionation and the origin of tholeiitic and alkaline magma types. Geochim. Cosmochim. Acta, 32: 1057—1086.

Hart, S.R., 1971. K, Rb, Cs, Cs, Sr and Ba contents and Sr isotope ratios of ocean floor basalts. Philos. Trans. R. Soc. London, Ser. A., 268: 573—582.
Hart, S.R., 1976. LIL — element geochemistry, Leg 34 basalts. In: Initial Reports of the Deep Sea Drilling Project, 35. U.S. Govt. Printing Office, Washington, D.C., in press.
Hedge, C.E. and Peterman, Z.E., 1970. The strontium isotopic composition of basalts fom the Gorda and Juan de Fuca Rises, Northeastern Pacific Ocean. Contrib. Mineral. Petrol., 27: 114—120.
Hedge, C.E., Watkins, N.D., Hildreth, R.A. and Doering, W.P., 1973. $^{87}Sr/^{86}Sr$ ratios in basalts from islands in the Indian Ocean. Earth Planet. Sci. Lett., 21: 29—34.
Hekinian, R., 1968. Rocks from the Mid-Oceanic Ridge in the Indian Ocean. Deep-Sea Res., 15: 195—213.
Hekinian, R., 1974a. Petrology of igneous rocks from Leg 22 in the northeastern Indian Ocean. In: Initial Reports of the Deep Sea Drilling Project, 22. U.S. Govt. Printing Office, Washington, D.C., pp. 413—447.
Hekinian, R., 1974b. Petrology of the Ninetyeast Ridge (Indian Ocean) compared to other aseismic ridges. Contrib. Mineral. Petrol., 43: 125—147.
Hekinian, R. and Thompson, G., 1976. Comparative geochemistry of volcanics from rift valleys, transform faults and aseismic ridges. Contrib. Mineral. Petrol., 57: 145—162.
Jakes, P. and White, A.J.R., 1972. Major and trace element abundances in volcanic rocks of orogenic areas. Geol. Soc. Am. Bull., 83: 29—40.
Kay, R.W. and Gast, P.W., 1973. The rare earth content and origin of alkali-rich basalts. J. Geol., 81: 653—682.
Kay, R., Hubbard, N.J. and Gast, P.W., 1970. Chemical characteristics and origin of oceanic ridge volcanic rocks. J. Geophys. Res., 75: 1585—1613.
Kempe, D.R.C., 1974. The petrology of the basalts, Leg 26. In: Initial Reports of the Deep Sea Drilling Project, 26. U.S. Govt. Printing Office, Washington, D.C., pp. 465—503.
Kempe, D.R.C. and Schilling, J.G., 1974. Discovery Tablemount basalt: Petrology and geochemistry. Contrib. Mineral. Petrol., 44: 101—115.
McKenzie, D.A. and Sclater, J.G., 1971. The evolution of the Indian Ocean since the late Cretaceous. Geophys. J., 25: 437—528.
Miyashiro, A. and Shido, F., 1975. Tholeiitic and calc-alkalic series in relation to the behaviours of titanium, vanadium, chromium and nickel. Am. J. Sci., 275: 265—277.
Morgan, W.J., 1971. Convection plumes in the lower mantle. Nature, 230: 42—43.
Morgan, W.T., 1972. Deep convection plumes and plate motions. Bull. Am. Assoc. Pet. Geol., 56: 203—213.
Peterman, Z.E. and Hedge, C.E., 1971. Related Sr isotopic and chemical variations in oceanic basalts. Geol. Soc. Am. Bull., 82: 498—500.
Philpotts, J.A., Schnetzler, C.C. and Hart, S.R., 1969. Submarine basalts: Some K, Rb, Sr, Ba, rare-earth, H_2O and CO_2 data bearing on their alteration, modification by plagioclase and possible source materials. Earth Planet. Sci. Lett., 7: 293—299.
Ramanan, S., 1976. Petrology, Geochemistry and Paleomagnetism of Volcanic Rocks North of Vajreswari. Thesis, I.I.T., Bombay.
Reddy, G.R., Rao, B.L., Pant, D.R. and Das, M.S., 1976. Neutron activiation analysis of 13 minor and trace elements in geological samples. J. Rad. Anal. Chem., 33: 39—51.
Schilling, J.G., 1971. Sea-floor evolution: rare earth evidence. Philos. Trans. R. Soc. London., Ser. A, 268: 663—706.
Schilling, J.G., 1973. Iceland mantle plume: Geochemical study of Reykjanes ridge. Nature, 214: 265—571.
Schilling, J.G. and Winchester, J.W., 1969. Rare earth contribution to the origin of Hawaiian lavas. Contrib. Mineral. Petrol., 23: 27—37.
Subbarao, K.V., 1972. The strontium isotopic composition of basalts from the East-Pacific Ocean. Contrib. Mineral. Petrol., 37: 111—120.
Subbarao, K.V. and Hedge, C.E., 1973. K, Rb, Sr, and Sr^{87}/Sr^{86} in rocks from the Mid-Indian Oceanic Ridge. Earth Planet. Sci. Lett., 18: 223—228.

Subbarao, K.V. and Hekinian, R., 1978. Alkali-enriched rocks from the central eastern Pacific Ocean. Mar. Geol., in press.
Subbarao, K.V., Viswanatha Reddy, V., Hekinian, R. and Chandrasekharam, D., 1976. Large ion lithophile elements and Sr and Pb isotopic variations in volcanic rocks from the Indian Ocean. Geol. Soc. Am. Mem., in press.
Taylor, S.R., Kaye, M., White, A.J.R., Duncan, A.R. and Ewart, A., 1969. Genetic Significance of Co, Cr, Ni, Se and V content of andesites. Geochim. Cosmichim. Acta, 33: 275—286.
Thompson, G., Bryan, W.B., Frey, F.A. and Sung, C.M., 1974. Petrology and geochemistry of basalts and related rocks from Sites 214, 215, 216, DSDP Leg 22, Indian Ocean. In: Initial Reports of the Deep Sea Drilling Project, 22. U.S. Govt. Printing Office, Washington, D.C., pp. 459—468.
Von der Borch, C.C., Sclater, J.G. et al., 1974. Initial Reports of the Deep Sea Drilling Project, 22. U.S. Govt. Printing Office, Washington, D.C.
Whitford, D.J., 1975. Strontium isotopic studies of the volcanic rocks of the Sunda arc, Indonesia, and their petrogenetic implications. Geochim. Cosmochim. Acta, 39: 1287—1302.
Wilson, J.T., 1965. Submarine fracture zones, aseismic ridges and the International Council of Scientific Unions line: proposed western margin of the east Pacific ridge. Nature, 207: 907—911.
Yoder, H.S. and Tilley, C.E., 1962. Origin of basaltic magmas: an experimental study of natural and synthetic rock systems. J. Petrol., 3: 342—532.
Zielinski, R.A., 1975. Trace element evaluation of a suite of rocks from Reunion Island, Indian Ocean. Geochim. Cosmochim. Acta, 39: 713—734.

Marine Geology, 26 (1978) 119—138

BASALTS AND RELATED ROCKS FROM DEEP-SEA DRILLING SITES IN THE CENTRAL AND EASTERN INDIAN OCEAN*

G. THOMPSON[1], W.B. BRYAN[1], F.A. FREY[2] and J.S. DICKEY Jr.[2]

[1] *Woods Hole Oceanographic Institution, Woods Hole, Mass. 02543 (U.S.A.)*
[2] *Massachusetts Institute of Technology, Cambridge, Mass. 02139 (U.S.A.)*

(Received March 28, 1977)

ABSTRACT

Thompson, G., Bryan, W.B., Frey, F.A. and Dickey Jr., J.S., 1978. Basalts and related rocks from deep-sea drilling sites in the Central and Eastern Indian Ocean. Mar. Geol., 26: 119—138.

Petrological and geochemical data are presented for basement rocks recovered from thirteen sites (211, 212, 213, 214, 215, 216, 253, 254, 256, 257, 259, 260 and 261) drilled on Legs 22, 26 and 27 of the Deep Sea Drilling Program. Basalts from Sites 212, 213, 253, 257, 259 and 261 have petrographic and geochemical characteristics of tholeiitic basalts from active spreading ridge axes in the major ocean basins.

Except for the lower basalts at Site 253, all the basement rocks from four Ninetyeast Ridge sites (214, 216, 253, 254) are similar to oceanic island tholeiitic sequences such as on Amsterdam—St. Paul Islands. The geochemical data for the Ninetyeast Ridge rocks are consistent with the development of this ridge as a hot-spot trace.

Basalts from deep ocean Sites 216 and 256 have geochemical characteristics more akin to tholeiitic basalts from spreading ridge axes close to major volcanic islands, i.e. they have high large ion lithophile element abundances relative to tholeiitic basalts found at spreading ridge axes away from islands.

Alkali olivine basalts were recovered from Site 211 and they contain abundant amphibole. They are related to the volcanism which created the Cocos—Keeling—Christmas shoal areas.

INTRODUCTION

We present summary data on the petrological and geochemical characteristics of basement rocks recovered from thirteen sites (211, 212, 213, 214, 215, 216, 253, 254, 257, 259, 260 and 261) drilled on Legs 22, 26 and 27 of the Deep Sea Drilling Project (Fig.1). Detailed petrography, major and trace element bulk rock compositions, and mineral analyses of individual samples have been presented in the Initial Reports for these legs and in Frey et al. (1977). The latter paper, in addition to presenting many new analyses, has evaluated the degree of correlation between petrological and geochemical

---

* WHOI Contribution No. 4026.

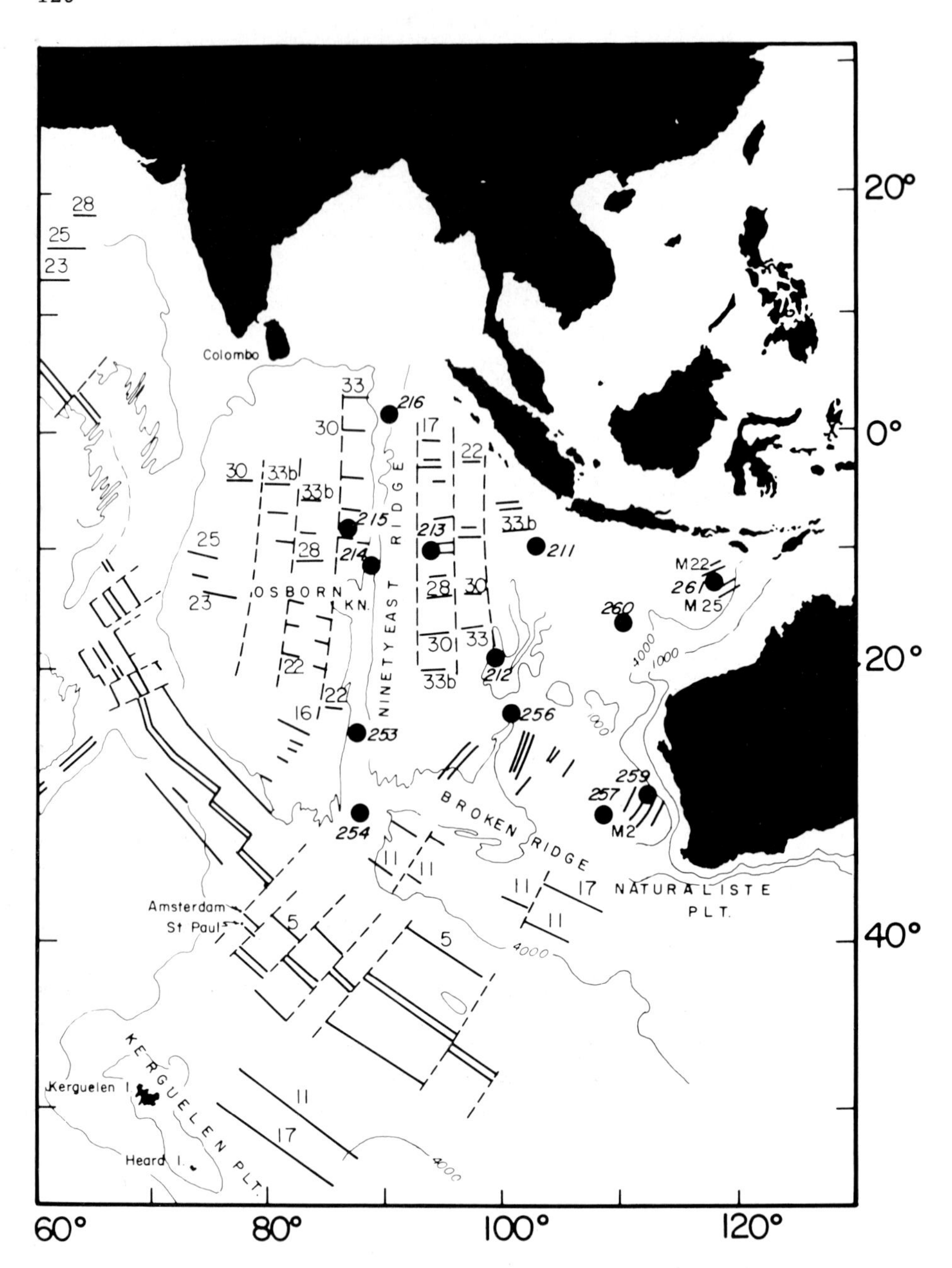

Fig.1. Location of DSDP sites on tectonic summary map of the eastern Indian Ocean (adapted from Luyendyk, 1977). Double solid lines are ridge axes, dashed lines are fracture zones, and short solid lines (some numbered) are magnetic anomalies.

features of basalts at each site with the postulated tectonic and geologic history of that site. In this paper we present only the salient petrologic and geochemical features of the major rock units at each site and indicate their similarities or contrasts with eruptive rocks obtained from similar tectonic environments in the Indian or other oceans.

CENTRAL INDIAN BASIN

Site 215 (8°07.30′S, 84°47.50′E)

The 25 m of basement penetrated at this site consist of a series of pillow flows with glassy margins. The textures and mineral paragenesis are characteristic of abyssal pillow basalts (Bryan, 1972). For example, plagioclase An_{68-73} is the common microphenocryst phase; however, the K_2O content of the plagioclase (0.15 wt.%) is higher than in the plagioclase of typical large-ion-lithophile (LIL)-element depleted basalts of spreading ridges. Fresh glass and whole rock major element analyses (Table I) indicate Site 215 basalts to be tholeiitic and similar to mid-ocean ridge basalts (MORB) except for unusually high K_2O (approximately 1.0 wt.%) and P_2O_5 (approximately 0.25 wt.%) contents. The trace element analyses (Table II) indicate that these basalts are much more enriched in LIL-elements such as Sr, Ba and Zr than MORB, and compared to chondrites, they are relatively enriched in light rare earth elements (LREE) — see Fig.2.

Thus, Site 215 basalts are not typical of basalts from the Mid-Indian Ocean ridge (or other mid-ocean ridge spreading centers). Tholeiitic basalts with similar trace element characteristics have been inferred to result from

TABLE I

Site 215: Major element composition (volatile—free, wt.%)*[1]

	Crystalline rock*[2] average	Glass average*[3]	Glass average*[4]	MIORB*[5]
SiO_2	50.4 ± 1.0	50.9 ± 0.5	50.67	50.67
Al_2O_3	16.8 ± 0.2	16.6 ± 0.1	16.93	16.60
FeO*	8.43 ± 0.21	8.11 ± 0.05	8.09	8.57
MnO	—	0.15 ± 0.1	—	0.16
MgO	6.48 ± 0.36	7.58 ± 0.05	7.08	8.02
CaO	10.95 ± 0.38	10.0 ± 0.34	10.59	11.58
Na_2O	3.17 ± 0.33	3.50 ± 0.12	3.04	2.75
K_2O	0.90 ± 0.06	0.97 ± 0.03	0.92	0.16
TiO_2	1.71 ± 0.02	1.56 ± 0.09	1.67	1.24
P_2O_5	0.31 ± 0.01	—	0.26	0.11
Total	99.15	99.37	99.25	99.86
FeO*/MgO	1.30	1.07	1.14	1.07

[1] In this, and all other tables, the reference sources making up the averages are identified. Details of individual samples making up these analyses can be found in the appropriate analytical tables for each site in Frey et al. (1977). FeO refers to total iron expressed as FeO.
*[2] Average and standard deviation of seven analyses (Hekinian, 1974; Thompson et al., 1974).
*[3] Average and standard deviation of thirteen glasses (Frey et al., 1977).
*[4] Average of eleven glasses (Melson et al., 1976).
*[5] Average Mid-Indian Ocean ridge basalt 8° S to 25° S (Engel and Fisher, 1975).

TABLE II

Site 215: Trace element composition (ppm)

	Crystalline rock average*[1]	Glass average*[2]	MIORB*[3]
B	4	<2	
Li	13	5	5.4 ± 4
Sc	47	39	49 ± 9
V	245	305	250 ± 60
Cr	250	310	360 ± 85
Co	50	50	35 ± 4
Ni	100	115	109 ± 26
Cu	65	58	93 ± 13
Sr	390	332	108 ± 26
Ba	450	415	17 ± 8
Y	36	43	35 ± 7
Zr	160	170	71 ± 18
La	16.2	15.1	2.67 ± 0.85
Ce	36	30	9.99 ± 6.12
Nd	20.1	14.5	8.79 ± 2.54
Sm	4.5	3.92	3.09 ± 0.86
Eu	1.5	1.45	1.15 ± 0.27
Tb	0.8	0.85	0.793 ± 0.214
Ho	—	0.81	1.03 ± 0.27
Yb	2.5	2.9	2.94 ± 0.86
Lu	0.51	0.44	0.517 ± 0.136

*[1] Average of five analyses (Thompson et al., 1974; Bougault, 1974; Frey et al., 1977).
*[2] Average of two analyses (Frey et al., 1977).
*[3] Mean and standard deviation for eighteen basalts from the Mid-Indian Ocean Ridge 8° S to 29° S (Engel and Fisher, 1975). The REE average is for five MIORB (Schilling, 1971).

mantle plumes (Schilling, 1973), but there are no unusual bathymetric or tectonic features suggestive of plume activity at Site 215. However, the relatively young basal sediment age at this site (59—60 · 10^6 years) compared to the magnetic anomaly age for this region of the Central Indian Basin is a puzzling feature (Sclater et al., 1974). Although only 240 km west of the Ninetyeast Ridge and 2—3° north of Site 214, basalts from Site 215 differ from the basalts of that ridge (Fig.2). In geochemical respects Site 215 basalts are similar to the LIL-element enriched tholeiitic basalts at Site 10, DSDP Leg 2 in the North Atlantic (Frey et al., 1974) which was inferred to be a non-spreading center, mid-plate eruption.

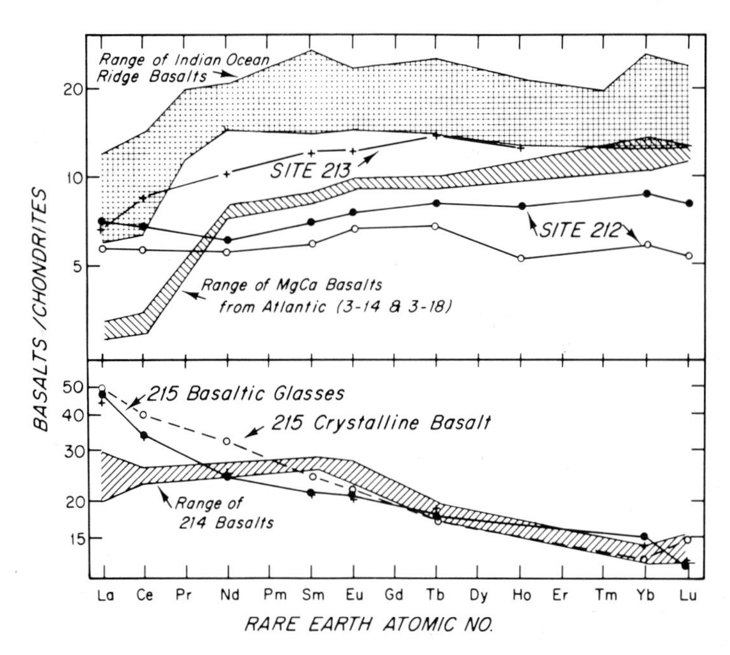

Fig.2. (Lower): REE abundances in two Site 215 basaltic glasses and a Site 215 crystalline basalt compared to range in Site 214 ferrotholeiites.
(Upper): REE abundances in two Site 212 metabasalts and a Site 213 basalt compared to REE range of five Mid-Indian Ocean Ridge basalts (Schilling, 1971) and high Mg, Ca ridge tholeiites from DSDP Leg 3, Sites 14 and 18 (Frey et al., 1974).

NINETYEAST RIDGE

Sites 214, 216, 253, 254 (11° 20.21′S, 88° 43.08′E; 1° 27.73′N, 90° 12.48′E; 24° 53′S, 87° 22′E; 30° 58′S, 87° 54′E)

Basalts from Sites 214 and 216 on the northern part of the Ninetyeast Ridge are vesicular and amygdaloidal in texture and thus are probably subaerial or shallow-marine eruptions. The typical trachytic groundmass texture distinguishes them from MORB which exhibit abundant quench features (Bryan, 1972). Plagioclase, pyroxene and magnetite are the principal mineral phases in these lavas; olivine is absent. The high abundance of typically euhedral magnetite (up to 10 vol.%) is unusual for sea-floor rocks.

At Site 216 only iron-rich basalt (ferrobasalt) was recovered, but at Site 214 similar ferrobasalts are overlain by differentiated rocks (54—58 wt.% SiO_2) termed oceanic andesites (Hekinian, 1974). These oceanic andesites are geochemically unlike those of island arc or continental margin andesites associated with subducting ocean lithosphere (Thompson et al., 1974). Despite significant compositional differences, the textures and mineral paragenesis of the ferrobasalts and oceanic andesites are similar. This similarity, in addition to their close association in time and space, implies a genetic relationship.

TABLE III

Major element composition (volatile–free wt.%) of basalts and oceanic andesites from the Ninetyeast Ridge and St. Paul—Amsterdam Islands

	Ferrobasalt*[1] Site 214	Ferrobasalt*[2] Site 216	Oceanic*[3] andesite Site 214	Ferrobasalt*[4] Site 254	Upper*[5] basalt Site 253	Lower*[6] basalt Site 253	St. Paul*[7] ferrobasalt	St. Paul*[8] andesite
SiO_2	48.1	49.5	56.9	47.66	52.20	47.20	48.39	55.4
Al_2O_3	14.9	13.5	15.9	15.37	18.30	16.30	12.6	12.6
FeO*	14.6	13.8	9.84	12.90	6.92	10.14	17.1	12.9
MgO	6.45	6.57	2.48	8.92	7.87	14.42	5.17	4.61
CaO	9.04	8.79	5.79	8.53	9.67	9.18	8.18	7.53
Na_2O	2.75	2.57	3.97	2.69	3.05	1.37	3.02	3.69
K_2O	0.37	0.90	1.50	0.32	0.32	0.28	0.46	0.97
TiO_2	2.35	2.75	1.45	2.12	0.98	0.69	3.58	2.05
P_2O_5	0.19	0.22	0.64	0.24	0.35	0.09	0.82	0.21
Total	98.75	98.60	98.47	98.75	99.71	99.67	99.32	99.96
FeO*/MgO	2.26	2.10	4.00	1.51	0.88	0.70	3.31	2.80

*[1] Average of six basalts (Thompson et al., 1974).
*[2] Average of seven basalts (Thompson et al., 1974).
*[3] Average of seven andesites (Hekinian, 1974).
*[4] Average of seven basalts (Kempe, 1974; Frey et al., 1977).
*[5] One analysis (Kempe, 1974).
*[6] Average of three basalts (Kempe, 1974; Frey et al., 1977).
*[7] One analysis (Girod et al., 1971).
*[8] One analysis (Girod et al., 1971).

TABLE IV

Trace element composition (ppm) of basalts and oceanic andesites from the Ninetyeast Ridge and St. Paul—Amsterdam Islands

	Ferrobasalt*[1] Site 214	Ferrobasalt*[2] Site 216	Oceanic*[3] andesite Site 214	Ferrobasalt*[4] basalt Site 254	Upper*[5] basalt Site 253	Lower*[6] basalt Site 253	St. Paul*[7] ferrobasalt	St. Paul*[8] andesite
Sc	46	42	16	42	50	51	35	25
V	525	445	39	342	210	209	395	250
Cr	38	45	5	469	345	749	55	125
Co	65	53	39	47	42	51	31	38
Ni	50	44	5	194	110	332	235	600
Sr	265	235	647	115	77	48	225	340
Ba	45	140	578	56	14	15	210	285
Y	26	31	65	50	32	25	67	95
Zr	120	159	252	156	100	21	330	355
La	8.4	13.2	32.7	9.9	9.2	3.7	19.6	28.3
Sm	5.1	5.0	11.3	4.8	3.1	2.2	7.26	8.9
Eu	1.7	1.45	3.17	1.63	1.2	0.78	2.30	2.46
Yb	3.0	3.6	4.39	3.3	2.9	2.6	3.6	4.8

*[1] Average of six basalts (Thompson et al., 1974; Frey and Sung, 1974).
*[2] Average of seven basalts (Thompson et al., 1974; Frey and Sung, 1974).
*[3] Average of five andesites (Thompson et al., 1974).
*[4] Average of thirteen basalts (Frey and Sung, 1974; Frey et al., 1977).
*[5] One analysis (Frey et al., 1977).
*[6] Average of five basalts (Frey et al., 1977).
*[7] One analysis (Frey et al., 1977).
*[8] One analysis (Frey et al., 1977).

The ferrobasalts of Sites 214 and 216 have similar compositional features (Table III): high absolute iron content (FeO* > 12.9 wt.%) with FeO*/MgO > 1.9 and TiO_2 > 2 wt.%. In addition, alkali metal content is low (K_2O < 1 wt.% and Na_2O < 3 wt.%). The tholeiitic nature of these basalts is also seen in their relatively low LIL-element abundances compared to alkalic basalts (Table IV). However, the low to moderate Ba, Zr and Sr contents and relative REE distribution with slight LREE enrichment relative to chondrites (Fig.3), clearly distinguish the Ninetyeast Ridge basalts from those of spreading ridges, including high-iron basalts such as those of the Juan de Fuca Ridge (Melson et al., 1976). These latter basalts have retained the LIL-element depleted characteristics of MORB (Kay et al., 1970). The Site 216 basalts are slightly more differentiated than those of Site 214. Qualitatively these differences can be accounted for by plagioclase—pyroxene—magnetite crystallization and fractionation (Frey et al., 1977).

The oceanic andesites of Site 214, relative to the ferrobasalts, are enriched in SiO_2, K_2O and Al_2O_3, but depleted in TiO_2, FeO*, MgO and CaO (Table III). In trace elements the oceanic andesites are very enriched in Sr, Ba, Zr and LREE (Fig.3), and depleted in Sc, V, Cr and Ni relative to the ferrobasalts (Table IV). Qualitatively the oceanic andesites may be derived from the ferrobasalts by clinopyroxene and magnetite separation (Frey et al., 1977).

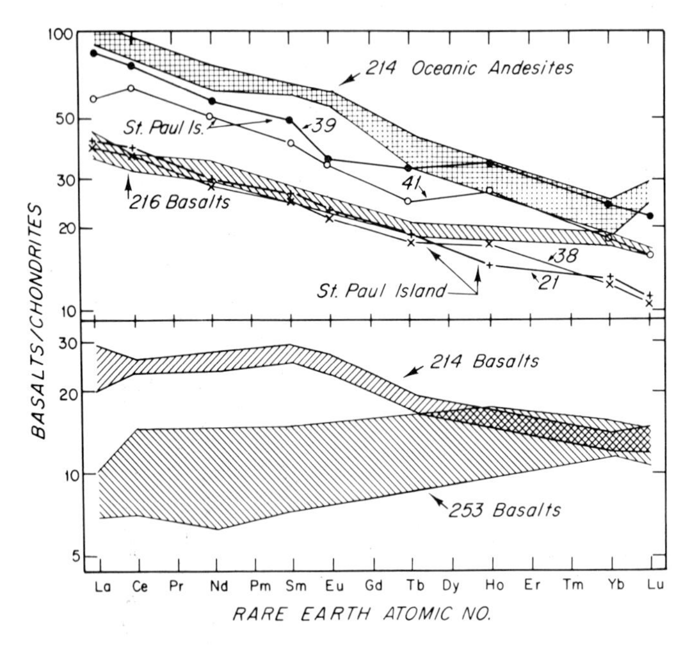

Fig.3. (Lower): REE abundance range in four Site 253 lower picritic tholeiites compared to range in Site 214 ferrotholeiites.
(Upper): REE abundances in four St. Paul Island rocks compared to range in Site 216 ferrotholeiites and Site 214 oceanic andesites.

Site 254 is located near the southern terminus of the Ninetyeast Ridge (Fig.1). Both coarse-grained and amygdaloidal basalts are found with pyroxene, plagioclase and some olivine as the major mineral phases. Although somewhat altered, the major element compositions (Table III) indicate they are high-iron basalts like those of Sites 214 and 216. The Site 254 basalts, however, have lower TiO_2 and FeO*/MgO than basalts of Sites 214 and 216. The trace element abundances (Table IV) show some similarities with Sites 214 and 216 basalts. Basalts from all three sites have moderate to high abundances (compared to MORB) of LIL-elements such as Ba, Zr, Hf and LREE. Certain major element characteristics (FeO*/MgO and TiO_2) and the compatible trace element abundances (Cr and Ni) imply that Site 254 basalt compositions are more primitive than Sites 214 and 216 basalts, although LIL-elements (e.g. La/Yb, Zr, Ba contents) are similar in all three sites. Thus it is unlikely that Sites 254, 214 and 216 basalts can be related in a simple petrogenetic model, although there are clearly many similar compositional features.

Site 253 basalts, on the western flank of the Ninetyeast Ridge some 6° north of Site 254, occur in two eruptive sequences: an upper scoriaceous flow in an ash sequence, and an underlying porphyritic olivine basalt. The upper basalts have low FeO*/MgO and TiO_2 abundances (Table III) compared to those of Sites 214, 216 and 254. However, in trace element composition (Table IV) the La/Yb and Hf content (> 3.5 ppm), and relative LREE enrichment, they show similarities and affinities with the basalt from other sites on the Ninetyeast Ridge.

The lower, olivine-rich flow has geochemical (Table III) and petrographic characteristics of a picritic basalt, i.e. an olivine cumulate (Frey and Sung, 1974). Trace element analyses (Table IV) show very low LIL-element abundances and LREE depletion characteristic of MORB (Fig.3). Thus, Site 253, although associated with the topographic high that forms the Ninetyeast Ridge, is anomalous in having two distinct eruptives, with the lower picritic sequence being from a mantle source akin to typical spreading ridge basalts and unlike the mantle source for basalts from other sites on the Ninetyeast Ridge.

The LIL-element enriched ferrobasalts of Sites 214, 216, 254 and possibly the upper flow of Site 253, appear to be distinctive for the Ninetyeast Ridge. They are similar in many aspects to iron-rich tholeiites (and associated oceanic andesites) in certain oceanic islands, e.g. Galapagos, Iceland, Faeroes, Amsterdam—St. Paul (Frey et al., 1977). The compositions of some representative samples of the latter volcanic complex are shown in Tables III and IV. Major element compositions, differentiation trends to ferrotholeiites and oceanic andesites, and high LIL-element abundances relative to tholeiitic basalts formed at spreading centers, all suggest the Ninetyeast Ridge rocks could be related to a volcanic source such as that now forming Amsterdam and St. Paul Islands. Therefore, the aseismic Ninetyeast Ridge could be a hot spot trace formed as the Indian plate moved over the Amsterdam—St. Paul Islands volcanic source (Frey et al., 1977).

WHARTON BASIN

Sites 212 and 213 (19°11.34′S, 99°17.84′E; 10°12.71′S, 93°53.77′E)

At Site 212 the basaltic basement consists of pillowed flows altered by halmyrolysis and hydrothermal processes. Chlorite—quartz—pumpellyite is a common mineral assemblage and these metabasalts range from zeolite to greenschist facies. Because of the altered nature, bulk rock alkali, Ca, Mg and Si concentrations vary widely probably as a result of chemical migration during hydrothermal alteration (Melson et al., 1968; Cann, 1969; Humphris, 1976). TiO_2 abundances, however, are consistently low (<0.7 wt.%), and this element has probably not been greatly affected by alteration (Cann, 1970; Humphris, 1976). Analyzed basalts (Table V) from this site that show little alteration, and particularly one chip of apparently fresh glass (Melson et al., 1976), have low contents of TiO_2 (0.61 wt.%), K_2O (0.05 wt.%), and P_2O_5 (0.03 wt.%); high MgO and CaO abundances and low FeO*/MgO. Apparently the Site 212 magma experienced little fractionation. Consistent with this interpretation are the high contents of compatible elements such as Cr and Ni and low LIL-element abundances such as Zr and REE (Table VI). Similar basaltic compositions occur at spreading ridge axes (Fig.2).

At Site 213 pillow basalts similar to those at Site 212 were cored. Alteration of these basalts is less than at Site 212 and is apparently restricted to

Table V

Sites 212 and 213: Major element compositions (volatile—free wt.%).

	Site 212		Site 213	
	Average crystalline*[1] rock	glass*[2]	average crystalline*[3] rock	average glass*[4]
SiO_2	51.23	51.33	48.83	49.66
Al_2O_3	18.75	15.60	18.09	15.92
FeO*	9.05	7.86	10.18	9.88
MnO	0.10	—	0.16	—
MgO	7.40	8.97	4.92	8.15
CaO	6.09	13.48	12.95	12.27
Na_2O	2.08	1.42	2.71	2.42
K_2O	3.26	0.05	0.58	0.06
TiO_2	0.62	0.61	1.02	1.07
P_2O_5	0.06	0.03	0.09	0.09
Total	99.64	99.35	99.53	99.52
FeO*/MgO	1.22	0.88	2.07	1.21

*[1] Average of four basalts (Hekinian, 1974; Frey et al., 1977).
*[2] One analysis (Melson et al., 1976).
*[3] Average of four basalts (Hekinian, 1974; Frey et al., 1977).
*[4] Average of six analyses (Melson et al., 1976).

TABLE VI

Sites 212 and 213: Trace element compositions (ppm)

	Site 212 crystalline rock*[1]	Site 213 crystalline rock*[2]
B	45	31
Li	33	18
Rb	62	29
Sc	46	43
V	215	217
Cr	1200	310
Co	39	45
Ni	140	83
Cu	56	86
Sr	80	97
Ba	79	13
Y	18	27
Zr	22	57
La	2.1	2.2
Ce	5.5	7.6
Nd	3.5	6.2
Sm	1.19	2.18
Eu	0.49	0.85
Tb	0.35	0.64
Ho	0.46	0.86
Yb	1.46	2.55
Lu	0.23	0.41

*[1] Average of two analyses (Frey et al., 1977).
*[2] Average of five analyses (Bougault, 1974; Frey et al., 1977; REE are only one analysis from Frey et al., 1977).

low-temperature weathering. Both major and trace element compositions (Tables V and VI) indicate geochemical characteristics of MORB, i.e. low K_2O, P_2O_5 contents and LIL-element depletion and relative depletion in LREE (Fig.2).

Site 211 and 256 (09°46.53′S, 102°41.95′E; 23°27.35′S, 100°46.46′E)

At Site 211 an intrusive 10 m thick diabase sill occurs 18 m above the basement basalt. The sill consists of phaneritic holocrystalline rock with plagioclase (An_{55}) and clinopyroxene (calcic, aluminous titan-augite) in ophitic and subophitic texture. The diabase is nepheline normative and its alkalic nature (Table VII) is reflected by high TiO_2 (> 2 wt.%) and $Na_2O + K_2O$ contents (>5 wt.%); high Sr and Ba abundances (Table VIII); and the marked LREE enrichment (Fig.4).

The basement basalts below the sill at Site 211 contain abundant

TABLE VII

Sites 211 and 256: Major element abundances (volatile—free wt.%)

	Site 211		Site 256
	Average, diabase sill*[1]	average, amphibole*[2] basalt	average basalt*[3]
SiO_2	46.86	46.71	50.36
Al_2O_3	15.64	18.41	13.34
FeO*	10.39	9.79	13.23
Mno	0.16	0.13	0.21
MgO	8.25	5.56	6.58
CaO	8.53	8.52	10.18
Na_2O	3.52	3.72	2.76
K_2O	2.25	2.41	0.25
TiO_2	2.33	2.79	2.36
P_2O_5	0.62	0.79	0.25
Total	98.55	98.33	99.52
FeO*/MgO	1.20	1.76	2.01

*[1] Average of five basalts (Hekinian, 1974; Frey et al., 1977).
*[2] Average of five basalts (Hekinian, 1974; Frey et al., 1977).
*[3] Average of seven basalts (Kempe, 1974; Frey et al., 1977).

amphibole (>50 vol.%) interpreted as primary and crystallizing from a hydrous magma (Hekinian, 1974). Major element compositions vary, reflecting some alteration, but in general (Table VII) are nepheline normative, have high TiO_2 (>2.1 wt.%) and total alkalies (>4.3 wt.%) typical of alkalic basalts. As in the overlying diabase sill, the basement basalt trace element contents (Table VIII) confirm the alkalic nature with high Sr, Ba, Zr, and LREE contents.

The basalts of Site 211, both sill and basement, are akin to the alkali olivine basalt suite characteristic of many oceanic islands. In particular, they are very similar to those on Christmas Island some 3° to the east. Thus, it is probable that Site 211 basalts are related to the Cocos—Keeling and Christmas shoal areas and not to a spreading ridge environment.

Site 256 in the southern Wharton Basin cored a series of flows containing plagioclase and clinopyroxene clusters with abundant magnetite. The plagioclase is zoned from An_{67} to An_{55}. Compositionally (Table VII) the basalts are FeO* and TiO_2-rich similar to the ferrobasalts of Sites 214 and 216 on the Ninetyeast Ridge. They also have high concentrations (Table VIII) of LIL-elements such as Sr, Ba, Y, Zr, and LREE (Fig.4). Thus, like the Ninetyeast Ridge basalts, those of Site 256 are ferrotholeiites similar to those of oceanic island sequences and unlike the high iron and titanium spreading center basalts which are depleted in LIL-elements. Rather than being formed at a spreading ridge the Site 256 basalts may be related to volcanism causing the series of topographic highs that strike from Broken Ridge to the vicinity of Site 256.

TABLE VIII

Sites 211 and 256: Trace element composition (ppm)

	Site 211		Site 256
	Average, diabase sill*[1]	average, amphibole*[2] basalt	average basalt*[3]
B	37	60	6
Li	37	83	7
Rb	47	70	—
Sc	21	18	46
V	173	164	451
Cr	293	63	108
Co	41	22	36
Ni	117	83	93
Cu	38	25	155
Sr	515	650	144
Ba	425	530	39
Y	27	37	45
Zr	160	300	159
La	47.5	32.8	9.8
Ce	74	82	31
Nd	28	37	17.5
Sm	6.12	6.76	4.69
Eu	1.95	2.19	1.52
Tb	1.0	1.0	1.0
Ho	1.1	0.85	1.3
Yb	2.6	2.8	3.3
Lu	0.29	0.37	0.55

*[1] Average of three basalts (Bougault, 1974; Frey et al., 1977; REE are only one analysis).
*[2] Average of four basalts (Bougault, 1974; Frey et al., 1977).
*[3] Average of eleven basalts (Kempe, 1974; Frey et al., 1977).

SOUTHEAST WHARTON BASIN

Sites 257 and 259 (39° 59′S, 108° 21′E; 29° 37′S, 112°42′E)

At Site 257 the basalt sequence alternates between medium and fine-grained vesicular basalt, and at least 7 or 8 flows are present (Kempe, 1974). Plagioclase, clinopyroxene and olivine have compositions typical of those in MORB (Frey et al., 1977). Three compositional groups (Table IX) have been recognized on the basis of FeO*/MgO and abundances of CaO, TiO_2, Cr, Ni, Zr, and REE (Table X). Group A is the least fractionated (FeO*/MgO < 0.93) and has high Cr and Ni contents and low Zr and heavy REE abundances. Group B is the most differentiated and Group C, the lowest unit, is intermediate in composition between A and B. The important feature of all Site 257 basalts is that they have the geochemical and petrographic characteristics of

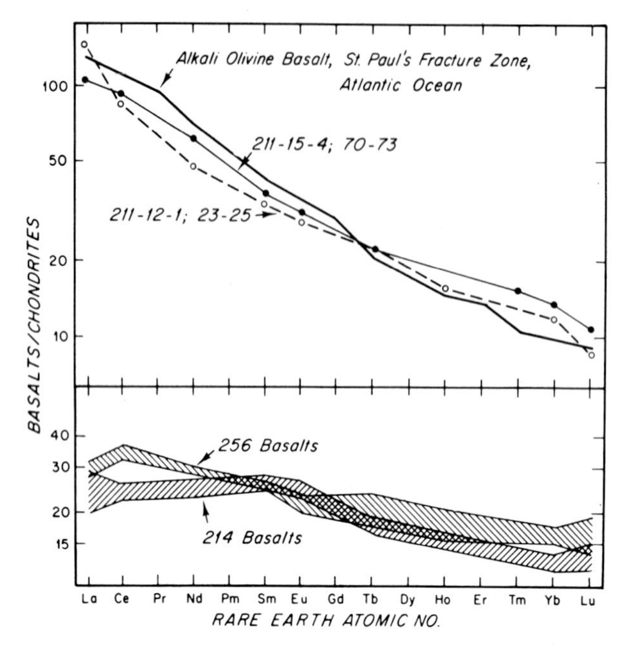

Fig.4. (Lower): REE abundance range in Site 256 ferrotholeiites compared to range in Site 214 ferrotholeiites.
(Upper): REE abundances in two Site 211 alkali olivine basalts and an alkali olivine basalt dredged from St. Paul's Fracture Zone in the Atlantic (Frey, 1970).

TABLE IX

Sites 257 and 259: Major element composition (volatile—free wt.%)

	Site 257			Site 259
	Unit A, average*[1]	Unit B, average*[2]	Unit C, average*[3]	average basalt*[4]
SiO_2	51.06	50.77	50.80	51.30
Al_2O_3	15.96	15.10	15.84	16.34
FeO*	7.72	10.74	8.73	9.70
MnO	0.25	0.22	0.19	0.17
MgO	9.37	7.67	8.37	7.80
CaO	11.95	11.32	12.60	8.95
Na_2O	2.33	2.26	2.12	2.51
K_2O	0.39	0.28	0.21	0.34
TiO_2	0.92	0.98	0.90	1.38
P_2O_5	0.17	0.10	0.09	0.15
Total	100.12	99.44	99.85	98.64
FeO*/MgO	0.82	1.40	1.04	1.24

*[1] Average of five basalts (Kempe, 1974; Frey et al., 1977).
*[2] Average of two basalts (Kempe, 1974; Frey et al., 1977).
*[3] Average of eight basalts (Kempe, 1974; Frey et al., 1977).
*[4] Average of four basalts (Robinson and Whitford, 1974).

TABLE X

Sites 257 and 259: Trace element composition (ppm)

	Site 257			Site 259
	Unit A, average*[1]	Unit B, average*[2]	Unit C, average*[3]	average basalt*[4]
B	25	30	18	27
Li	20	8	10	9
Sc	47	44	47	—
V	306	305	282	265
Cr	379	162	290	150
Co	43	43	39	38
Ni	146	105	122	58
Cu	44	78	64	52
Sr	103	70	78	85
Ba	72	18	19	8
Y	38	39	32	34
Zr	28	45	39	71
La	1.44	1.58	1.41	1.8
Ce	6.8	6.0	4.9	6.7
Nd	4.5	5.8	4.9	5.3
Sm	1.68	2.09	1.66	2.0
Eu	0.64	0.79	0.71	0.76
Tb	0.46	0.60	0.36	0.53
Ho	0.68	0.93	0.79	0.95
Yb	2.0	2.9	2.5	2.8
Lu	0.43	0.48	0.42	0.43

*[1] Average of five basalts.
*[2] Average of two basalts.
*[3] Average of twelve basalts.
(Kempe, 1974; Frey et al., 1977. The B, Sc, Co, Zr, and REE are only from analyses by Frey et al.)
*[4] Average of three basalts (Robinson and Whitford, 1974; Frey et al., 1977).

LIL-element depleted tholeiites typical of spreading centers (Fig.5).

At Site 259 moderately altered basalts contain plagioclase (An_{55-60}) and rare clinopyroxene. Geochemically (Tables IX and X) they are characterized by low concentrations of Sr, Ba, and Zr and relative depletion in LREE (Fig.5). Compositionally they are very similar to the Group B basalts of Site 257. Basalts from Sites 257 and 259 are thus consistent with an origin at a common spreading axis as inferred by Markl (1974).

NORTHEAST WHARTON BASIN

Sites 260 and 261 (16°9'S, 110°18'E; 12°57'S, 117°54'E)

At Site 260 the basalt is believed to be a sill (Robinson and Whitford, 1974). The basalts are fresh, medium-grained and sparsely porphyritic. Plagioclase, clinopyroxene and magnetite are present. Compositionally

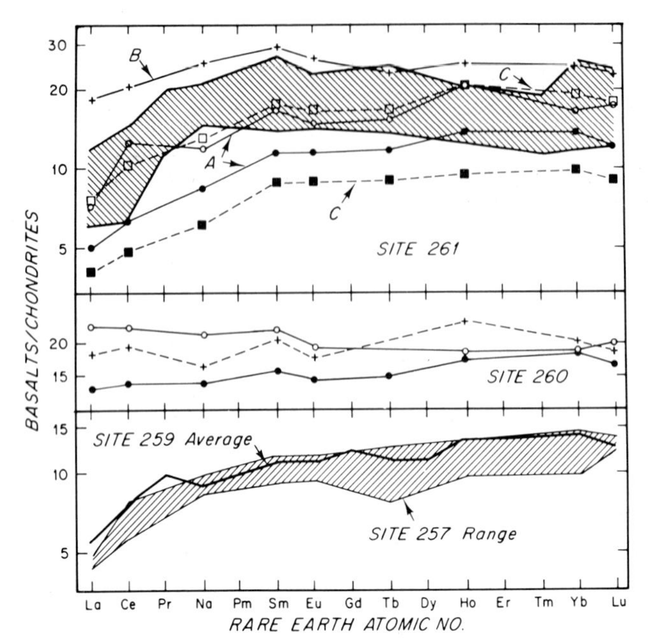

Fig.5. (Lower): REE abundance average in Site 259 basalts (Robinson and Whitford, 1974) compared to REE range in Site 257 basalts.
(Middle): REE abundances in Site 260 basalts.
(Upper): REE abundances in Site 261 basalts (Unit C ■—■; Unit B +—+; Unit A ○—○) compared to range for five MIORB (Schilling, 1971).

(Tables XI and XII) they are quite variable with relatively high TiO_2 and LIL-element contents compared to MORB, i.e. relatively high Ba, and Zr contents and a nearly chondritic REE distribution (Fig.5). In petrographical and geochemical respects, these basalts are intermediate between LIL-element depleted MORB and island tholeiites.

At Site 261 three separate eruptive units were penetrated (Robinson and Whitford, 1974). Unit A, the upper unit, is a 10 m thick coarse-grained sill containing plagioclase, augite and olivine. Unit B, the middle sequence, is a 3-m section of fine-grained highly altered basalt containing plagioclase, clinopyroxene, ilmenite and olivine. Unit C is the basal pillow basalt—breccia complex containing plagioclase, clinopyroxene and minor olivine.

The upper sill, Unit A, compositionally (Table XI) is very similar to MORB and the basal pillows (Unit C) of this site. Both units have low LIL-element abundances (Table XII) and relative LREE depletion (Fig.5). The altered basalts of Unit B are marked by very high TiO_2 contents ($>$ 3.3 wt.%), and high abundances of FeO*, P_2O_5, total alkalies and LIL-elements. Unit B basalts differ compositionally from units A and C but are more like the basalts of Site 260. Thus the petrographic and geochemical data for Sites 260 and 261 basalts are consistent with origin at or near a spreading ridge axis. However, the presence of sills and of some basaltic compositions that differ slightly from typical MORB, indicate either complex volcanism with

TABLE XI

Sites 260 and 261: Major element compositions (volatile—free wt.%)

	Site 260	Site 261		
	Average basalt*[1]	Unit A, average*[2]	Unit B, average*[3]	Unit C, average*[4]
SiO_2	51.43	50.51	49.32	49.99
Al_2O_3	15.73	14.74	15.68	15.32
FeO*	11.31	10.06	10.99	9.22
MnO	0.25	0.20	0.29	0.22
MgO	6.92	7.69	7.33	7.52
CaO	8.02	11.39	6.72	12.04
Na_2O	3.36	2.39	3.48	2.72
K_2O	0.24	0.40	0.88	0.55
TiO_2	1.79	1.34	3.40	1.09
P_2O_5	0.13	0.13	0.31	0.12
Total	99.18	99.85	98.49	98.79
FeO*/MgO	1.63	1.31	1.50	1.23

*[1] Average of three basalts (Robinson and Whitford, 1974; Frey et al., 1977).
*[2] Average of three basalts (Robinson and Whitford, 1974; Frey et al., 1977).
*[3] Average of two basalts (Robinson and Whitford, 1974).
*[4] Average of three basalts (Robinson and Whitford, 1974; Frey et al., 1977).

magmas intruded some distance from an active ridge crest, and/or possibly extensive clinopyroxene ± spinel fractionation of a LIL-element depleted tholeiitic magma (Frey et al., 1977).

CONCLUSIONS

Based on petrologic and geochemical studies of hard rocks from the scattered DSDP sampling sites in the central and eastern Indian Ocean we make the following conclusions:

(1) Basalts from Sites 212, 213, 253, 257, 259 and 261 have petrographic and geochemical features within the ranges established by dredged tholeiitic basalts from active spreading ridge axes in the Atlantic, Pacific and Western Indian Oceans. Although there are important exceptions, the majority of basalts forming layer 2 of the central and eastern Indian Ocean were apparently formed at spreading ridge axes. This conclusion is consistent with tectonic models based on geophysical data.

(2) Except for the lower basalts at Site 253, all the igneous rocks recovered on the four Ninetyeast Ridge Sites (214, 216, 253, 254) are akin to oceanic island tholeiitic sequences, such as Amsterdam—St. Paul, and unlike basalts formed at spreading ridge axes away from islands. The geochemical data for Ninetyeast Ridge rocks are consistent with the development of the Ninetyeast Ridge as a hot-spot trace.

(3) Basalts from the deep ocean sites (215 and 256) have geochemical

TABLE XII

Site 260 and 261: Trace element compositions (ppm)

	Site 260	Site 261		
	Average basalt*[1]	Unit A, average*[2]	Unit B, average*[3]	Unit C, average*[4]
B	13	25	48	20
Li	12	12	17	8
Sc	61	52	—	52
V	385	317	460	335
Cr	81	157	8	150
Co	72	36	17	41
Ni	166	86	56	83
Cu	67	80	62	45
Sr	107	84	78	115
Ba	45	3	23	8
Y	46	43	65	48
Zr	128	76	185	85
La	6.0	2.0	6.0	1.9
Ce	16	8.3	18	6.6
Nd	10.3	6.1	15	5.7
Sm	3.5	2.61	5.3	2.4
Eu	1.17	0.91	1.8	0.89
Tb	0.7	0.64	1.1	0.61
Ho	1.4	1.2	1.8	1.09
Yb	3.8	3.1	5.1	2.9
Lu	0.63	0.53	0.79	0.46

*[1] Average of three basalts (Robinson and Whitford, 1974; Frey et al., 1977).
*[2] Average of two basalts (Robinson and Whitford, 1974; Frey et al., 1977).
*[3] Average of two basalts (Robinson and Whitford, 1974; Frey et al., 1977).
*[4] Average of two basalts (Robinson and Whitford, 1974; Frey et al., 1977).

features typical of oceanic island tholeiites and basalts dredged from spreading ridge axes near islands. In terms of Schilling's model (1973) such basalts have formed as a result of mantle plumes. From the vicinity of Site 256 southwest to Broken Ridge there are a series of bathymetric highs; thus it is possible that Site 256 basalts formed as a result of volcanism causing this anomalous bathymetry.

Although Site 215 is located near the Ninetyeast Ridge it is bathymetrically distinct from the ridge and geochemically unlike the nearest Ninetyeast Ridge basalts studied (Site 214). As at Leg 2, Site 10 in the Atlantic, there is no satisfactory understanding of why Site 215 tholeiitic basalts have high LIL-element abundances relative to tholeiitic basalts formed at spreading ridge axes away from islands.

(4) Alkali olivine basalts were recovered at Site 211, and contain abundant amphibole, a mineral not usually encountered in deep sea basalts. This site is

near Christmas Island and Site 211 basalts must be related to volcanism which has created the Cocos—Keeling—Christmas shoal areas.

(5) Basalts from Site 260 and the middle unit of Site 261 have geochemical features intermediate between island tholeiites and LIL-element depleted tholeiites formed at spreading ridges. At Site 261 these intermediate basalts are stratigraphically between units of LIL-element depleted basalt. More detailed study of this site and its stratigraphic succession will be useful in understanding the petrogenesis of contrasted ocean floor basalt types.

ACKNOWLEDGEMENTS

D. Bankston, J. Guertler, R. Houghton, S. Roy and M. Sulanowski assisted with sample preparation and analysis. DSDP samples were supplied through the assistance of the National Science Foundation, and this research was supported by NSF grants DES 74-00268 to the Woods Hole Oceanographic Institution and DES 74-00147 to the Massachusetts Institute of Technology.

REFERENCES

Bougault, H., 1974. Distribution of first series transition metals in rocks recovered during DSDP Leg 24 in the Northeastern Indian Ocean. In: C.C. von der Borch, J.G. Sclater et al., Initial Reports of the Deep Sea Drilling Project, 22. U.S. Govt. Printing Office, Washington, D.C., pp.449—457.

Bryan, W.B., 1972. Morphology of quench crystals in submarine basalts. J. Geophys. Res., 77: 5812—5819.

Cann, J.R., 1969. Spilites from the Carlsberg Ridge, Indian Ocean, J. Petrol., 10: 1—19.

Cann, J.R., 1970. Rb, Sr, Y, Zr and Nb in some ocean floor basaltic rocks. Earth Planet. Sci. Lett., 10: 7—11.

Engel, C.G. and Fisher, R.L., 1975. Granitic to ultramafic rock complexes of the Indian Ocean Ridge System, western Indian Ocean. Geol. Soc. Am. Bull., 86: 1553—1578.

Frey, F.A., 1970. Rare earth and potassium abundances in St. Paul's rocks. Earth Planet. Sci. Lett., 7: 351—360.

Frey, F.A. and Sung, C.M., 1974. Geochemical results for basalts from Sites 253 and 254. In: T.A. Davies, B.P. Luyendyk et al., Initial Reports of the Deep Sea Drilling Project, 26. U.S. Govt. Printing Office, Washington, D.C., pp.567—572.

Frey, F.A., Bryan, W.B. and Thompson, G., 1974. Atlantic Ocean floor: Geochemistry of basalts from Legs 2 and 3 of the Deep Sea Drilling Project. J. Geophys. Res., 79: 5507—5527.

Frey, F.A., Dickey, J.S., Thompson, G. and Bryan, W.B., 1977. Eastern Indian Ocean DSDP sites: Correlations between petrography, geochemistry and tectonic setting. In: J.R. Heirtzler and J.G. Sclater (Editors), A Synthesis of Deep Sea Drilling in the Indian Ocean. U.S. Govt. Printing Office, Washington, D.C., in press.

Girod, M., Camus, G. and Vialette, Y., 1971. Discovery of tholeiitic rocks at St. Paul Island (Indian Ocean). Contrib. Mineral. Petrol., 33: 108—117.

Hekinian, R., 1974. Petrology of igneous rocks from Leg 22 in the Northeastern Indian Ocean. In: C.C. von der Borch, J.G. Sclater et al., Initial Reports of the Deep Sea Drilling Project 22. U.S. Govt. Printing Office, Washington, D.C., pp.413—447.

Humphris, S., 1976. The Hydrothermal Alteration of Oceanic Basalts by Seawater. Thesis, Woods Hole Oceanographic Institution, Woods Hole, Mass., 246 pp.

Kay, R., Hubbard, N. and Gast, P., 1970. Chemical characteristics and origin of oceanic ridge volcanic rocks. J. Geophys. Res., 75: 1585—1613.
Kempe, D.R.C., 1974. The petrology of the basalts, Leg 26. In: T.A. Davies, B.P. B.P. Luyendyk et al., Initial Reports of the Deep Sea Drilling Project, 26. U.S. Govt. Printing Office, Washington, D.C., pp.465—503.
Luyendyk, B.P., 1977. Deep sea drilling on the Ninetyeast Ridge: Synthesis and a tectonic model. In: J.R. Heirtzler and J.G. Sclater (Editors), A Synthesis of Deep Sea Drilling in the Indian Ocean. U.S. Govt. Printing Office, Washington, D.C., in press.
Markl, R.G., 1974. Bathymetric map of the Eastern Indian Ocean. In: T.A. Davies, B.P. B.P. Luyendyk et al., Initial Reports of the Deep Sea Drilling Project, 26. U.S. Govt. Printing Office, Washington, D.C., pp.967—968.
Melson, W.G., Thompson, G. and Van Andel, T.H., 1968. Volcanism and metamorphism in the Mid-Atlantic Ridge, 22°N latitude. J. Geophys. Res., 73: 5925—5941.
Melson, W.G., Vallier, T.L., Wright, T.L., Byerly, G. and Nelen, J., 1976. Chemical diversity of abyssal volcanic glass erupted along Pacific, Atlantic and Indian Ocean sea floor spreading centers. In: Geophysics of the Pacific Ocean Basin and its Margins. Geophys. Monogr. 19: 351—368 (Am. Geophys. Union, Washington, D.C.).
Robinson, P.T. and Whitford, D.J., 1974. Basalts from the Eastern Indian Ocean, DSDP Leg 27. In: J.J. Veevers, J.R. Heirtzler et al., Initial Reports of the Deep Sea Drilling Project, 27. U.S. Govt. Printing Office, Washington, D.C., pp.551—559.
Schilling, J.G., 1971. Sea floor evolution: rare-earth evidence. Philos. Trans. R. Soc. London, Ser. A, 268: 663—703.
Schilling, J.G., 1973. Iceland mantle plume: geochemical evidence along Reykjanes Ridge. Nature, 242: 565—571.
Sclater, J.R., Von der Borch, C.C. et al., 1974. Regional synthesis of the deep sea drilling results from Leg 22 in the Eastern Indian Ocean. In: C.C. von der Borch, J.G. Sclater et al., Initial Reports of the Deep Sea Drilling Project, 22. U.S. Govt. Printing Office, Washington, D.C., pp.815—831.
Thompson, G., Bryan, W.B., Frey, F.A. and Sung, C.M., 1974. Petrology and geochemistry of basalts and related rocks from Sites 214, 215, 216, DSDP Leg 22, Indian Ocean. In: C.C. von der Borch, J.G. Sclater et al., Initial Reports of the Deep Sea Drilling Project, 22. U.S. Govt. Printing Office, Washington, D.C., pp.459—468.

Marine Geology, 26 (1978) 139—175

LATE CENOZOIC BENTHONIC FORAMINIFERA OF THE NINETYEAST RIDGE (INDIAN OCEAN)

ESTEBAN BOLTOVSKOY

Museo Argentino de Ciencias Naturales "Bernardino Rivadavia", Buenos Aires (Argentina) and Consejo Nacional de Investigaciones Científicas y Técnicas, Buenos Aires (Argentina)

(Received March 28, 1977)

ABSTRACT

Boltovskoy, E., 1978. Late Cenozoic benthonic foraminifera of the Ninetyeast Ridge (Indian Ocean). Mar. Geol., 26: 139—175.

A study was made of the benthonic foraminifera recovered from 242 sediment samples (ranging from Upper Oligocene through Quaternary) at five DSDP sites (214, 216, 217, 253 and 254) located on the Ninetyeast Ridge (Indian Ocean). The sites lie at depths varying between 1,253 and 3,010 m.

The sediments cored are believed to be undisturbed and do not contain reworked specimens. The stratigraphic subdivision was made on the basis of previous planktonic foraminiferal and other microplankton studies.

In total 170 species were identified. About twenty species turned out to be unidentifiable. Two species were put in *nomenclatura aperta*. No new taxon has been described.

The fauna studied exhibits strong dominance. The following species were found consistently and often frequently in the whole sequence: *Anomalina globulosa*, *Cibicides kullenbergi*, *Epistominella exigua*, *Eponides bradyi*, *Gyroidina lamarckiana*, s.l., *G. soldanii*, *Oridorsalis umbonatus*, *Pullenia osloensis*, *Stilostomella* ex gr. *S. lepidula* and *Uvigerina proboscidea*, s.l.

Very few species were found to be stratigraphically diagnostic. It was impossible to detect the Oligocene/Lower Miocene boundary by means of benthonic foraminifera. The Lower Miocene/Middle Miocene boundary is characterized by: (1) the dramatic decrease of the *Planulina marialana gigas* population and increase of the *Cibicides wuellerstorfi* population; and (2) the first appearance of a very typical foraminifer of uncertain origin named conditionally "*Bulava indica*". The Middle Miocene/Upper Miocene boundary is characterized by the first appearance of *Bolivina globulosa* and the extinction of *Bolivinopsis cubensis*, *Bulimina jarvisi* and *Cassidulina caudriae*. At the Upper Miocene/Pliocene boundary "*Bulava indica*", *Cassidulina cuneata*, *C. subglobosa horizontalis*, and *Rectuvigerina royoi* disappear.

Two evolutionary trends: *Planulina marialana gigas* → *Cibicides wuellerstorfi* and *Cassidulina cuneata* → *Cassidulina subglobosa* were observed.

INTRODUCTION

In my work on the Neogene deep-water benthonic foraminifera of the Indian Ocean (Boltovskoy, in prep.) it is shown that ten of the fifteen DSDP Indian Ocean sites contained redeposited specimens which at some sites were distributed throughout the whole length of the cores. Previous authors who studied the planktonic components of the same samples did not realize this and therefore considered their paleontological and stratigraphic determinations as reliable. Unfortunately, in many cases it is rather difficult to decide if redeposited specimens should be considered as displaced from the sediments of the same age, but located in the shallower water, or reworked from older strata. In the latter case, however, all the material is unreliable and thus the age determination of the samples discussed, as well as all the stratigraphic conclusions drawn on them, are questionable.

Five sites were drilled in the Indian Ocean by the Deep Sea Drilling Project on the Ninetyeast Ridge. The examination of these sites revealed that all of them contain well-represented Neogene sequences in situ, presumably without any reworked elements. This circumstance, as well as the fact that all the sites are located in the bathyal zone above the $CaCO_3$ compensation depth and contain virtually continuous sequences, prompted me to dedicate a special study to the benthonic Late Cenozoic fauna of the Ninetyeast Ridge.

MATERIAL AND METHODS

For the present report I had at my disposal the material from the five sites drilled on the Ninetyeast Ridge. The Ninetyeast Ridge is a long narrow (≅ 100 km) crustal elevation rising about 2,000 m above the sea bottom. It is located more or less parallel to the 90°E meridian, approximately between latitudes 9°N and 32°S. The sides of the ridge are characterized by steep slopes. It belongs tectonically to the Indian Plate. The geological description of this ridge is to be found in Davis et al. (1974) and Von der Borch et al. (1974). The sites are plotted on Fig.1.

TABLE I

Location of sites, their depth and number of samples

Leg	Site	Latitude	Longitude	Depth (m)	Number of samples
22	214	11°20.21′S	88°43.08′E	1655	56
22	216	01°27.73′N	90°12.48′E	2237	20
22	217	08°55.57′N	90°32.33′E	3010	15
26	253	24°52.65′S	87°21.91′E	1962	76
26	254	30°58.15′S	87°53.72′E	1253	75

Each site contains a complete Neogene sequence underlain by Upper Oligocene sediments. The geological age of the samples was established based on previous studies of planktonic foraminifera and other microfossils (Boltovskoy, 1974; McGowran, 1974).

The benthonic foraminifera of Sites 214, 216 and 217 were not studied prior to this work. Accordingly, about 300 specimens were picked at random

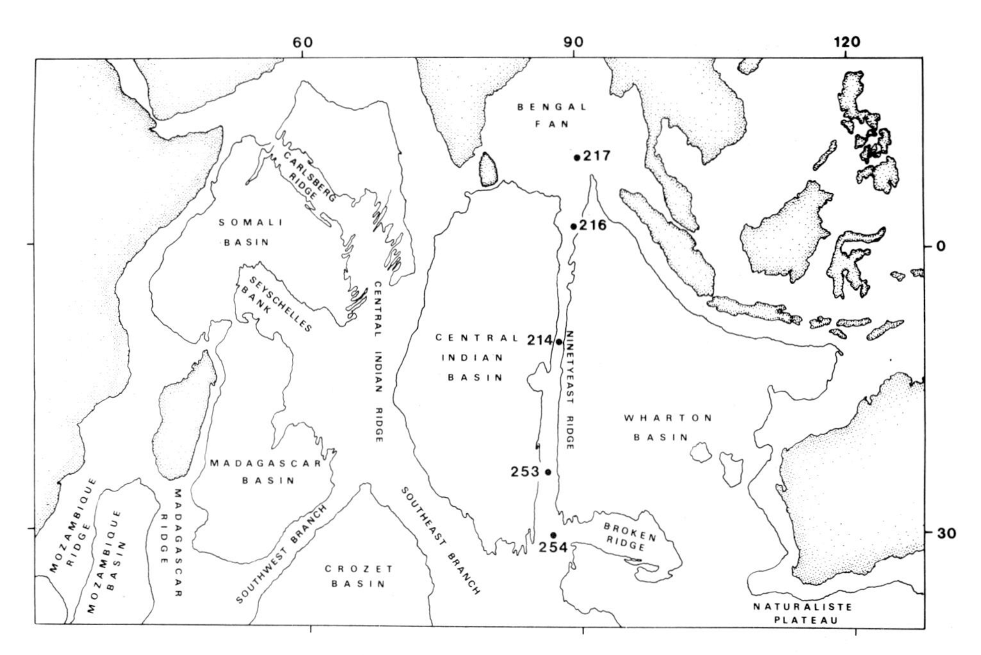

Fig.1. Physiographic features of the Indian Ocean, as defined by the 400-m contour (after Davies et al., 1974) and the location of the sites studied.

from each sample of the sites mentioned. The benthonic foraminifera of Sites 253 and 254 were previously studied by Boltovskoy (in prep.). At that time about 150 specimens were picked from each sample. However, for the present report all the samples of Sites 253 and 254 were re-checked, some additional picking was carried out, and some small improvements in taxonomic identification were realized.

PROBLEMS OF TAXONOMICAL IDENTIFICATION

Correct taxonomical identification of specimens is of prime importance in zoological, paleontological and stratigraphical studies.

Unfortunately, the taxonomical confusion in the foraminiferal household, especially that of benthonic foraminifera, is enormous, and is constantly growing with time (Boltovskoy, 1965).

Several species have been found in the present material which do not fit the species known in the literature and some of them may be new. However, I preferred to put them in *nomenclatura aperta* if they were represented by a considerable number of specimens and were easily distinguished, and leave them undescribed and uncited if they occurred as rare, isolated tests. No new species have been established in this report because I feel that it is quite probable that studies based on richer material would show that these "new species" are nothing other than ecological variants of existing ones.

On the other hand, many of my species very well fitted the description of several taxa, and to choose the right one without the original material posed a problem of tradition. I am quite sure that a lot of names in use are synonyms. To put a name in a synonymy, however, is also a rather drastic step, and to carry it, it is necessary to base it on good modern descriptions and figures or, better still, by comparison with type collections. Not having such material, I had to eliminate this step also. Therefore, I am not quite sure that all foraminiferal names which I use in this report have real zoological value and I am fully aware that many of them would be named differently by different colleagues.

In light of all this confusion, I think that the best way to describe the relevant foraminifera is by illustration, initial citation and, if necessary, short remarks. For this reason care was taken to illustrate all the species recovered, and in the case of variable species, also their main variants. This procedure was especially important for the present report, because, as far as I know, benthonic deep-water foraminifera from the DSDP cores of the Indian Ocean as yet have not been illustrated.

The following example demonstrates the taxonomical confusion in the benthonic foraminiferal classification and the importance of initial citations and figures. In 1974 Burmistrova published a study on benthonic foraminifera collected at depths between 2,200 and 4,500 m in the area located approximately between 20° N and 10° S latitude and 77° E and 87° E longitude This area is part of both the Ninetyeast Ridge and an area immediately adjacent to it. Burmistrova cited in her work somewhat more than 30 species. In the present study I found twice as many in the Quaternary sediments. This is quite understandable because I had many more samples at my disposal. It would be logical to expect that our lists of species found would be similar or at least somewhat alike. However, only four species are found on both lists. Unfortunately, even this cannot be verified because for each of them we used different generic names and Burmistrova does not give either primary citations and figures or even the author's name of some species. This simply illustrates the comparative problems involved.

GENERAL CHARACTERISTICS OF THE BENTHONIC FAUNA

Rich, well-preserved Upper Oligocene through Quaternary benthonic foraminiferal assemblages were recovered at all the sites on the Ninetyeast Ridge.

The total number of species which were identified down to species level is 170. I was not able to identify about twenty species; two of them are figured, described, and put in *nomenclatura aperta.* The identification of eleven species was uncertain and, consequently, their names are accompanied by ?, "cf." or "aff.". Four species were interpreted in a broad sense and "ex gr." or "s.l." are used.

The assemblages found at different sites are rather similar. The majority of the most common and consistent species which make up the faunas are the same at all the five sites examined. Some difference can be seen only in their relative abundance and, in some cases, in insignificant morphological differences.

Another characteristic feature of these assemblages is that they exhibit strong dominance. Usually, in each sample about 25—45 species were identified (minimum 16, maximum 48). However, 1—3 species were dominant and composed the bulk of the specimens.

TABLE II

Highest percentage of dominant species observed in one sample

Unit	1 species	3 species
Quaternary	55%	70%
Pliocene	60%	78%
Upper Miocene	48%	69%
Middle Miocene	47%	60%
Lower Miocene	68%	80%
Upper Oligocene	50%	73%

Many of the remaining species were represented by 1—2 specimens only.

The following species were found consistently (and in many sections frequently or even abundantly) in the whole Upper Oligocene—Quaternary sequence: *Anomalina globulosa*, *Cibicides kullenbergi*, *Epistominella exigua*, *Eponides bradyi*, *Gyroidina lamarckiana*, s.l., *G. soldanii*, *Oridorsalis umbonatus*, *Pullenia osloensis*, *Stilostomella* ex gr. *S. lepidula* and *Uvigerina proboscidea*, s.l.

The following species were recovered also throughout the whole sequence studied, but not consistently, and on the average in smaller numbers of specimens; however, in some sections some of them were frequent and even abundant: *Astrononion umbilicatulum*, *Bolivina pusilla*, *B.* cf. *B. thalmanni*, *Bulimina rostrata*, *Cibicides bradyi*, *Eggerella bradyi*, *Eponides polius*, *Gyroidina umbonata*, *Karreriella bradyi*, *Osangularia culter*, *Pullenia subcarinata quinqueloba*, *Stilostomella* cf. *S. annulifera* and *Uvigerina peregrina*.

Four species, *Dentalina communis*, *Laticarinina pauperata*, *Robulus rotulatus*, s.l., and *Sphaeroidina bulloides* turned out to be consistent at all the sites from the Oligocene up to the Quaternary, but their occurrence practically everywhere was rare or very rare.

As all the foraminifera cited above occurred throughout the whole Oligocene—Quaternary sequence, naturally, they cannot be used as guide fossils.

Of the species with common or frequent occurrence only the following have a restricted time range which makes them suitable for stratigraphic purposes. *Cassidulina subglobosa subglobosa* and *Cibicides wuellerstorfi*, which were abundant or frequent in Quaternary—Middle Miocene sections, but absent or quite rare in older deposits. *Cassidulina cuneata* and *Planulina marialana gigas*, on the contrary, were frequent in the Oligocene and the lower half of the Miocene, rare in the upper half of the Miocene, and absent in the Pliocene and Quaternary. *Bolivina globulosa* was absent in Oligocene—Middle Miocene and common in the Upper Miocene—Quaternary sequences. *Discorbis subvilardeboanus* was recovered in considerable quantity in Oligocene—Lower Miocene deposits at Sites 214 and 216, but missing in younger sections.

The remaining species with limited time ranges were not recovered from all sites and, in addition, usually were rare or very rare. Thus, to find them it was necessary to check large amounts of material. However several of them, as will be shown below, can serve as guide fossils.

GENERAL RANGE CHART OF SELECTED SPECIES AND MAJOR STRATIGRAPHIC BOUNDARIES

Table III shows the vertical distribution of 48 selected benthonic species. The foraminifera selected are those dominant for the whole sequence under discussion and those which, thanks to their limited time range and reasonable abundance, can be used (at least in assemblages) as guide fossils. It should be noted that the abundance signs are a very rough average of the data obtained from the five sites and therefore are not precise.

The vertical distribution of the selected species listed was found to be very similar at all the sites. This is one more proof that the sites of the Ninetyeast Ridge contain an undisturbed, homogeneous Late Cenozoic sequence without heterogeneous elements. On the other hand, the ranges of almost all species were found to be approximately the same as those established by their authors. Naturally, there were several discrepancies but not important ones. The greatest discrepancy is probably in *Pleurostomella obtusa*, a species described in the Lower Cretaceous sediments of France but found as high as the Pliocene of the Ninetyeast Ridge. However, this is not the first case of a Cretaceous species occurring in Pliocene sediments.

Below is a description of the faunal changes at the main stratigraphic boundaries.

The Oligocene/Lower Miocene boundary

As we can see on the range chart (Table III), no change in benthonic foraminiferal fauna takes place at this boundary.

The Lower Miocene/Middle Miocene boundary

It is difficult to locate this boundary by means of benthonic foraminifera, and this location is uncertain. "*Bulava indica*" has not been found at any sites of the Indian Ocean in the Lower Miocene sediments. Other species which can help in locating this boundary are *Cibicides notocenicus* and *Discorbis subvilardeboanus*. The former is abundant in the Lower Miocene and was found as isolated specimens in the Middle and Upper Miocene. *Discorbis subvilardeboanus* was found in the Middle Miocene. Unfortunately, both species have a restricted geographic distribution: *Cibicides notocenicus* was found at three sites and *Discorbis subvilardeboanus* at two sites. In addition, *Planulina marialana gigas*: *Cibicides wuellerstorfi* ratio can also be used to detect this boundary as the former species strongly predominates over the latter in the Lower Miocene, but in the Middle Miocene *C. wuellerstorfi* is much more abundant than *P. marialana gigas*. The presence or absence of some other species cannot be taken into account because these species are found as isolated specimens.

The Middle Miocene/Upper Miocene boundary

At this boundary *Bolivina globulosa* appears as a rather frequent species and *Bolivinopsis cubensis*, *Bulimina jarvisi*, *Cassidulina caudriae*, *Kyphopyxa* sp. "A" and *Pleurostomella* cf. *P. praegerontica* all disappear.

The Upper Miocene/Pliocene boundary

This boundary is rather well pronounced because here the following typical Miocene fossils disappear: "*Bulava indica*", *Bulimina macilenta*, *Cassidulina cuneata*, *C. subglobosa horizontalis*, *Saracenaria latifrons jamaicensis* and *Stilostomella tuckerae*. The best Upper Miocene indicators from this group are: *Bulimina macilenta*, *Cassidulina subglobosa horizontalis* and especially "*Bulava indica*". As for *Rectuvigerina royoi*, this species can also be considered as a guide fossil for Upper Miocene deposits, because only two specimens were found in the lowermost Pliocene sediments, only at Site 254.

The Pliocene/Quaternary boundary

This boundary is very indistinct and uncertain. *Heronallenia* sp. "A", various *Orthomorphina*, *Pleurostomella* and *Vulvulina pennatula* were recovered in the Pliocene sequence of the Ninetyeast Ridge but were absent in the younger sediments. However, the occurrence of these species is low, they are not found at all the sites, and in addition, the majority of these species are well known in other areas in the Quaternary sediments.

BENTHONIC FORAMINIFERA AS INDICATOR FOSSILS

By means of planktonic foraminifera it was possible not only to subdivide the Pliocene and Miocene deposits of the Ninetyeast Ridge into lower, middle and upper parts, but in many cases even to define zones (Boltovskoy, 1974; McGowran, 1974).

This is not the case with the benthonic foraminiferal faunas. They do not show any significant change in their vertical distribution. An attempt to adjust the range chart based on benthonic foraminifera to the divisions made by means of planktonic ones, as we have seen above, was successful only at some boundaries.

Other shortcomings of the use of benthonic species in the stratigraphy of deep sea sediments are the difficulties in picking out an adequate number of specimens, and the difficulty of correlating the results with those of other researchers. Benthonic foraminifera usually comprise about 0.1—0.5% of the total foraminiferal assemblage (Douglas, 1973). This means that much more time and energy must be used to pick a given quantity of benthonic foraminifera tests than for the same quantity of planktonic ones.

Correlations of the stratigraphic columns based on benthonic foraminifera range charts is very difficult because of the enormous confusion which exists in benthonic foraminiferal taxonomy; this problem was discussed above.

The statement by Douglas (1973) that many bathyal species have a wide geographic distribution and a limited stratigraphic range is correct as far as its first part is concerned, i.e. they have really wide geographic distribution. However, their stratigraphic ranges in the majority of the cases are also wide. In addition, many of these species are rare numerically. For these reasons deep sea benthonic foraminifera cannot be considered as good time indicators. It seems, therefore, that the conclusion of Douglas (1973, p.607) that benthonic foraminifera ". . . offer an excellent basis for biostratigraphic correlation" is somewhat exaggerated. My experience with the Neogene benthonic deep-water foraminifera of the whole Indian Ocean (Boltovskoy, in prep.), those of the Ninetyeast Ridge (this report) and those of the South Atlantic Ocean (in preparation) allow me to draw a less optimistic conclusion. They can be used stratigraphically, but only for the subdivision of major units, and even in these cases the results obtained are not always satisfactory.

In connection with this it is of interest to compare the results obtained by Douglas (1973) in his study of benthonic foraminifera in the central North Pacific and those of the present study. I take this paper by Douglas because it is one of the most detailed recent studies of Late Cenozoic deep-water benthonic foraminifera based on the DSDP cores. Douglas listed 54 species for the Upper Oligocene—Quaternary section which are common; some of them are considered by Douglas as stratigraphic indicators. Table III of this study shows the stratigraphic range of 48 common and/or stratigraphically valuable species found also in the Upper Oligocene—Quaternary sequence of the Ninetyeast Ridge. Some of these species are also interpreted as indicators.

Only ten species, namely; *Anomalina globulosa* (*Anomalinoides globulosa*, according to Douglas), *Bolivinopsis cubensis*, *Bulimina jarvisi*, *B. macilenta*, *Cassidulina subglobosa*, *Cibicides wuellerstorfi* (*Cibicidoides wuellerstorfi*, according to Douglas), *Oridorsalis umbonatus*, *Pullenia subcarinata quinqueloba* (*P. quinqueloba*, according to Douglas), *Uvigerina peregrina* and *Uvigerina proboscidea* are in both lists.

A comparison of figures and bibliographical data given by Douglas and those of this study reveals that the real number of common species is somewhat greater. It is evident, for example, that *Bulimina rostrata* of this study is *B. alazanensis* in the interpretation of Douglas. A comparison of the collections would reveal several additional synonyms. However, the final result of the comparison of both faunas is not very promising as far as the

TABLE III

The vertical distribution of selected benthonic species
A broken line signifies very rare or rare occurrence
A solid line signifies regular, frequent or abundant occurrence

Species	QUATERNARY	PLIOCENE	U. MIOCENE	M. MIOCENE	L. MIOCENE	U. OLIGOCENE
Anomalina globulosa	solid	solid	broken	broken	solid	broken
Astrononion umbilicatulum	solid	solid	solid	solid	solid	solid
Bolivina globulosa	solid	solid	solid			
Bolivina pusilla	broken	broken	broken	broken	broken	solid
Bolivina cf. B. thalmanni	broken	solid	solid	broken	broken	broken
Bolivinopsis cubensis				broken	broken	broken
"Bulava indica"			solid	broken		
Bulimina jarvisi				broken	broken	solid
Bulimina macilenta			broken	broken	broken	broken
Bulimina rostrata	broken	solid	broken	broken	broken	broken
Cassidulina caudriae				broken	broken	broken
Cassidulina cuneata			broken	solid	solid	solid
Cassidulina subglobosa horizontalis			broken	broken	broken	broken
Cassidulina subglobosa subglobosa	solid	solid	solid	broken	broken	broken
Cibicides bradyi	broken	broken	broken	broken	solid	broken
Cibicides kullenbergi	solid	solid	solid	solid	solid	solid
Cibicides notocenicus			broken	broken	solid	solid
Cibicides wuellerstorfi	solid	solid	solid	solid	broken	
Discorbis subvilardeboanus					broken	broken
Eggerella bradyi	broken	broken	broken	broken	broken	broken
Epistominella exigua	solid	solid	solid	solid	solid	solid
Eponides bradyi	solid	solid	solid	solid	broken	broken
Eponides polius	broken	broken	broken	broken	broken	broken
Gyroidina lamarckiana, s.l.	solid	solid	solid	solid	solid	solid
Gyroidina soldanii	broken	broken	solid	solid	solid	solid
Gyroidina umbonata	broken	broken	broken	broken	broken	broken
Heronallenia sp. "A"		broken	broken	broken	broken	broken
Karreriella bradyi	broken	broken	broken	broken	broken	broken
Kyphopyxa sp. "A"				broken	broken	broken
Oridorsalis umbonatus	solid	solid	solid	solid	solid	solid
Orthomorphina cf. O. antillea		broken	broken	solid	solid	solid
Orthomorphina himerensis		broken	broken	broken	broken	broken
Osangularia culter	broken	broken	solid	broken	broken	broken
Planulina marialana gigas				broken	solid	solid
Pleurostomella dominicana		broken	broken	broken	broken	broken
Pleurostomella obtusa		broken	broken	solid	broken	broken
Pleurostomella cf. P. praegerontica				broken	broken	broken
Pullenia osloensis	solid	solid	solid	solid	broken	solid
Pullenia quadriloba			broken	broken	broken	broken
Pullenia subcarinata quinqueloba	broken	solid	broken	broken	broken	broken
Rectuvigerina royoi			broken	broken	broken	
Saracenaria latifrons jamaicensis			broken	broken	broken	broken
Stilostomella cf. S. annulifera	broken	solid	solid	solid	broken	broken
Stilostomella ex gr. S. lepidula	broken	solid	solid	solid	solid	solid
Stilostomella tuckerae			broken	broken	broken	broken
Uvigerina peregrina	solid	broken	broken	broken	broken	broken
Uvigerina proboscidea, s.l.	solid	solid	solid	broken	solid	broken
Vulvulina pennatula		broken	broken	broken	broken	solid

possibility of the wide use of benthonic foraminifera for the world stratigraphy of deep-water sediments is concerned. Although many Late Cenozoic benthonic deep-water foraminiferal species have worldwide geographic distribution, many of them have either too wide a stratigraphic range, or this range was the result of different processes, according to Douglas and this study (Table IV).

Nevertheless, the role of benthonic foraminifera in a study of deep sea core sediments can be very important in two instances:

(1) When calcareous fauna suffers dissolution, benthonic species are more resistant, and are thus still identifiable when planktonic ones are completely dissolved.

(2) For the recognition of redeposited material, planktonic foraminifera can be labelled as heterogeneous to the sediment, either if they are of quite different age, or if they were evidently transported. As, however, different benthonic species are typical of different depths, it is quite easy to conclude that not all the material is in situ if a mixture of benthonic dwellers from different bathymetric zones is found.

TABLE IV

Time range of the benthonic foraminiferal species found both in the Indian and North Pacific Oceans

Species	Sections					
	Olig.	L.Mio.	M.Mio.	U.Mio.	Plio.	Quat.
Anomalina globulosa	--- . . .	--- . . .	---	---	---	---
Bolivinopsis cubensis	--- . . .	---	---			
Bulimina rostrata (= *B. alazanensis*)	--- . . .	--- . . .	---			
Bulimina jarvisi	--- . . .	--- . . .	---			
Bulimina macilenta	---	---	---	---		
Cassidulina subglobosa subglobosa	--- . . .	--- . . .	--- . . .	--- . . .	--- . . .	--- . . .
Cibicides wuellerstorfi		---	--- . . .	--- . . .	--- . . .	--- . . .
Oridorsalis umbonatus	--- . . .	--- . . .	--- . . .	--- . . .	--- . . .	--- . . .
Pullenia subcarinata quinqueloba	--- . . .	--- . . .	--- . . .	--- . . .	--- . . .	--- . . .
Uvigerina peregrina	---	---	---	--- . . .	--- . . .	--- . . .
Uvigerina proboscidea	---	---	---	---	--- . . .	--- . . .

Broken line = Indian Ocean, according to the present study.
Dotted line = Pacific Ocean, according to Douglas (1973).

EVOLUTIONARY TRENDS

(1) *Planulina marialana gigas/Cibicides wuellerstorfi trend*

One of the most abundant species in the Quaternary through the Upper Miocene is *Cibicides wuellerstorfi*. It is still rather frequent in the Middle Miocene, but rare or very rare in the Lower Miocene, and absent in the Oligocene.

A quite contrary vertical distribution was observed in the same strata in the *Planulina marialana gigas* population. This foraminifer is abundant in the Oligocene and Lower Miocene, rare or very rare in the Middle Miocene, and absent in younger strata.

P. marialana gigas is represented in the Oligocene by very typical specimens. In the Lower Miocene section many specimens became thinner, more plano-convex and the aperture changes locality from its peripheral location to the evolute side. These morphological changes draw those specimens very close to *C. wuellerstorfi*. In the Middle Miocene the above-mentioned tendency increases to such an extent that only rare specimens still preserve the features of typical *P. marialana gigas*, whereas the number of transitional forms as well as those of typical *C. wuellerstorfi* grows considerably. As mentioned above, *P. marialana gigas* disappears in the Upper Miocene, being replaced by *C. wuellerstorfi*.

This process of gradually transforming *P. marialana gigas* to *C. wuellerstorfi* is very well observed at almost all the sites throughout the Miocene, but the "critical point" is the Lower Miocene/Middle Miocene boundary. Approximately at this boundary, *C. wuellerstorfi* begins to predominate over *P. marialana gigas* up-section. An exception, however, is Site 254, at which in the Lower Miocene section no *P. marialana gigas* specimens were found at all.

(2) *Cassidulina cuneata/Cassidulina subglobosa, s.l. trend*

This trend is more complicated and not completely understood.

Cassidulina subglobosa subglobosa is abundant or frequent in the Quaternary, Pliocene and Upper Miocene, but it begins to decline numerically in the lower part of the Upper Miocene and this tendency continues downward into the Middle and Lower Miocene sections. Nevertheless, this species still occurs, although rarely, in the Lowermost Miocene sediments, and atypical isolated specimens are found even in the uppermost Oligocene sequence. The morphology of this species also changes downward. In the Upper Miocene the aperture of several specimens becomes large, gaping, semicircular or sometimes almost straight and at an angle of about 90° from the axis. The whole test becomes irregularly wedge-shaped, the height:width ratio of the test decreases, small grooves appear in the apertural area, and the quantity and location of chambers approaches that of *Ehrenbergina*. Not all these changes are evident in every specimen. Specimens with a large semicircular aperture and irregular wedge-shaped form approach *Cassidulina cuneata*, and tests with the aperture located at an angle to the test axis are *Cassidulina subglobosa horizontalis*.

In the Middle Miocene sediments the former tendency is so well pronounced in many specimens that typical *C. cuneata* specimens are frequent. At the same time specimens appear in which the height:width ratio decreases considerably, which makes them resemble *C. carandelli*.

In the Lower Miocene *C. cuneata* is considerably more abundant than *C. subglobosa subglobosa* and *C. subglobosa horizontalis*.

In the Upper Oligocene *C. cuneata* is abundant or frequent, *C. subglobosa subglobosa* and *C. subglobosa horizontalis* are very rare.

On the basis of both the above described vertical distribution of the species discussed and their morphological changes, the following tentative evolutionary trend can be suggested.

In Oligocene time three offshoots developed from *C. cuneata: C. subglobosa subglobosa*, *C. subglobosa horizontalis*, and probably at the end of this period *C. carandelli. C. subglobosa subglobosa* turned out to be the most resistant and long-living offshoot and it is living today. However, during Early and Middle Miocene time it shared the same habitat with other offshoots of *C. cuneata* and was not very abundant. During Late Miocene time *C. subglobosa horizontalis* and *C. cuneata* were very rare and at the beginning of the Pliocene both became extinct; *C. subglobosa subglobosa* then occupied their niche and became abundant.

This evolutionary trend is very general and further studies are necessary to clear up several details. The evolutionary position of *C. carandelli* is probably the most questionable.

SYSTEMATIC PALEONTOLOGY

The faunal reference list contains the preferred modern name of the species, followed by its original name. The data on geographic location, stratigraphic occurrence and abundance of the species in the area studied directly follow the original species name and are grouped according to geological ages. In parentheses after the geological age, the number(s) of the site(s) is(are) given, separated by a comma from a symbol indicating the abundance of the species, viz.: *R* = very rare or rare; *F* = frequent; *C* = common or abundant.

Allomorphina pacifica Cushman and Todd, 1949, Cushman Lab. Foraminiferal Res., 25(3): p.68, pl.12, f.6—9. L.Mio.(214,R); M.Mio.(214,216,R); U.Mio.(214,R); Plio.(214,R).

PLATE I

1. *Allomorphina pacifica* Cushman and Todd; × 65; site 214.
2, 3. *Ammosphaeroidina sphaeroidiniformis* (Brady); × 80 (fig.2), × 35 (fig.3); site 253.
4. *Angulogerina occidentalis* (Cushman); × 115; site 254.
5—8. *Anomalina globulosa* Chapman and Parr; × 35; site 253.
9, 10. *Anomalina alazanensis* Nuttall; × 45; site 216.
11. *Astacolus crepidulus* (Fichtel and Moll); × 75; site 254.
12. *Astacolus increscens* (Reuss); × 30; site 254.
13. *Astrononion stelligerum* (d'Orbigny); × 65; site 214.
14. *Bolivina anastomosa* Finlay; × 75; site 216.
15. *Astrononion umbilicatulum* Uchio; × 75; site 254.
16—18. *Bolivina pusilla* Schwager; × 50 (fig.16), × 40 (figs.17,18); site 253.
19. *Bolivina globulosa* Cushman; × 75; site 253.
20—23. *Bolivina silvestrina* Cushman; × 60; site 254.
24—27. *Bolivina* cf. *B. thalmanni* Renz; × 65 (fig.24), × 70 (fig.25), × 60 (figs.26,27); site 253 (figs.24,27), site 214 (figs.25,26).
28. *Bolivina* cf. *B. villalverniensis* Martinis; × 50; site 253.
29. *Bolivina subspinescens* Cushman; × 100; site 216.
30—32. *Bolivinopsis cubensis* (Cushman and Bermúdez); × 30; site 217 (fig.30), site 254 (figs.31,32).
33—35. "*Bulava indica*"—Boltovskoy (1977); × 50 (fig.33), × 75 (fig.34), × 100 (fig.35); site 253 (figs.33,35), site 216 (fig.34).
36, 37. *Bulimina inflata* Seguenza; × 50; site 254.
38, 39. *Bulimina macilenta* Cushman and Parker; × 40; site 253.
40, 41. *Bulimina miolaevis* Finlay; × 35; site 254.

PLATE I

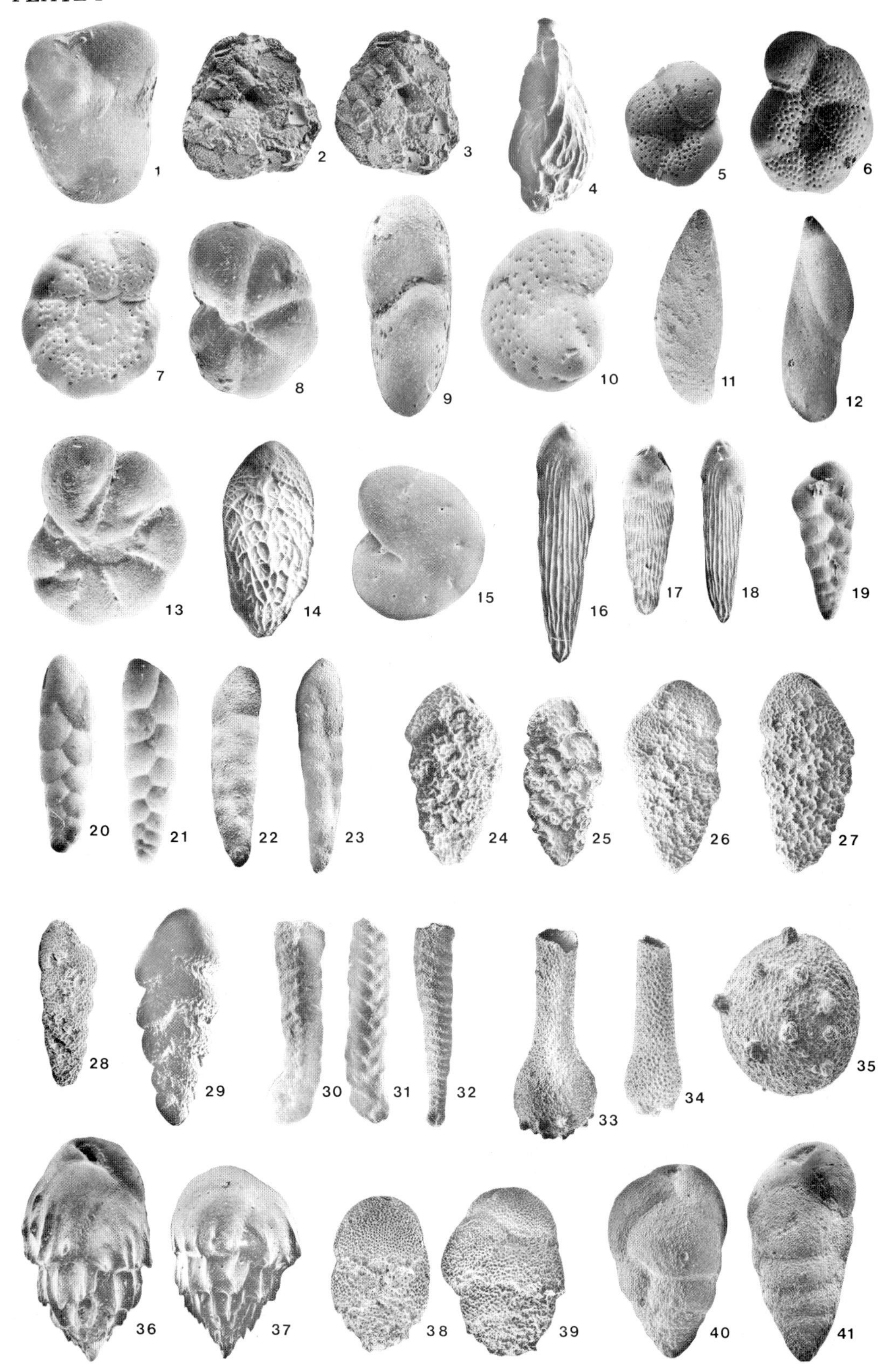

1
2
3
4
5
6
7
8
9
10
11
12
13
14
15
16
17
18
19
20
21
22
23
24
25
26
27
28
29
30
31
32
33
34
35
36
37
38
39
40
41

Ammosphaeroidina sphaeroidiniformis (Brady) = *Haplophragmium sphaeroidiniformis* Brady, 1884, "Challenger" Exped. Rep., Zool., 9: p.313. M.Mio.—Plio.(254,R).
Angulogerina occidentalis (Cushman) = *Uvigerina occidentalis* Cushman, 1923, U.S.Natl. Mus., Bull., 104(4): p.169, pl.5, f.4,5. L.Mio.—M.Mio.(254,R).
Anomalina alazanensis Nuttall, 1932, J. Paleontol., 6(1): p.31, pl.8, f.5—7. U.Mio.(216,R).
Anomalina globulosa Chapman and Parr, 1937, Australas.Antarct. Exped., C, 1: p.117, pl.9, f.27. Olig.—Quat.(214,216,217,253,254,R—F).
Astacolus crepidulus (Fichtel and Moll) = *Nautilus crepidula* Fichtel and Moll, 1798, Test.Microsc., p.107, pl.19, f.g—i. Olig.—M.Mio.(253,254,R); Plio.(254,R); Quat.(253, R).
Astacolus increscens (Reuss) = *Cristellaria* (*Cristellaria*) *increscens* Reuss, 1863. K.Akad. Wiss., Math.-Naturwiss, Cl., S.—B., 48(1): p.50, pl.4, f.47,48. Olig.—M.Mio.(214,216, 253,254,R).
Astrononion stelligerum (d'Orbigny) = *Nonionina stelligera* d'Orbigny, 1839, In: Barker, Webb and Berthelot, Hist. Nat. Iles Canaries, 2(2), Foraminifères,: p.128, pl.3, f.1,2. Plio.(214,R).
Astrononion umbilicatulum Uchio, 1952, Jpn. Assoc. Pet. Tech. J., 17(1): p.36, tf.1. Olig.—U.Mio.(214,216,217,C; 217,R); Plio.(214,216,217,254,R—F); Quat.(214,216, C; 217,R).
Bolivina anastomosa Finlay, 1939, R.Soc.N.Z., Trans. Proc., 69(part 3): p.320, pl.27, f.75—77,103,111. L.Mio.(254,F); Plio.—U.Mio.(216,C).
Bolivina globulosa Cushman, 1933, Cushman Lab.Foraminiferal.Res., Contrib., 9(part 4): p.80, pl.8, f.9. U.Mio.(214,216,217,C; 253,R); Plio.(214,216,217,F—C); Quat.(214,C).
Bolivina pusilla Schwager, 1866, "Novara" Exped., Geol., 2: p.254, pl.7, fig.101. Olig.—L.Mio.(214,216,217,R—F); M.Mio.—U.Mio.(214,216,217,253,254,R—F); Plio.(214, 253,254,R—F);Quat.(214,216,253,R—F).

PLATE II

1—3. *Bulimina jarvisi* Cushman and Parker; × 30; site 253.
4—7. *Bulimina rostrata* Brady; × 35; site 253.
8. *Bulimina translucens* Parker; × 100; site 217.
9—11. *Bulimina semicostata* Nuttall; × 100; site 253.
12. *Buliminella sculpturata* Keijzer; × 40; site 253.
13, 14. *Buliminella carteri* Bhatia; × 65 (fig.13), × 50 (fig.14); site 214 (fig.13), site 253 (fig.14).
15. *Carpenteria balaniformis* Gray; × 20; site 254.
16. *Cassidella bradyi* (Cushman); × 25; site 214.
17. *Cassidulina carandelli* Colom; × 35; site 254.
18. *Cassidulina carapitana* Hedberg; × 65; site 214.
19. *Cassidulina crassa* d'Orbigny, forma minima Boltovskoy; × 135; site 217.
20—22. *Cassidulina coudriae* Cushman and Stainforth; × 40 (figs.20,21), × 60 (fig.22); site 214 (fig.22), site 216 (figs.20,21).
23, 24. *Cassidulina cuneata* Finlay; × 50 (fig.23), × 35 (fig.24); site 217 (fig.23), site 254 (fig.24).
25, 26. *Cassidulina elegans* Sidebottom; × 35 (fig.25), × 75 (fig.26); site 214.
27. *Cassidulina laevigata* d'Orbigny; × 75; site 217.
28. *Cassidulina limbata* Cushman and Hughes; × 50; site 214.
29. *Cassidulina minuta* Cushman; × 115; site 216.
30. *Cassidulina monicana* Cushman and Kleinpell; × 50; site 254.
31. *Cassidulina oblonga* Reuss; × 115; site 253.
32. *Cassidulina subglobosa horizontalis* Cushman and Renz; × 30; site 253.
33. *Cassidulina subglobosa producta* Chapman and Parr; × 20; site 253.
34. *Cassidulina subglobosa subglobosa* Brady; × 30; site 214.

PLATE II

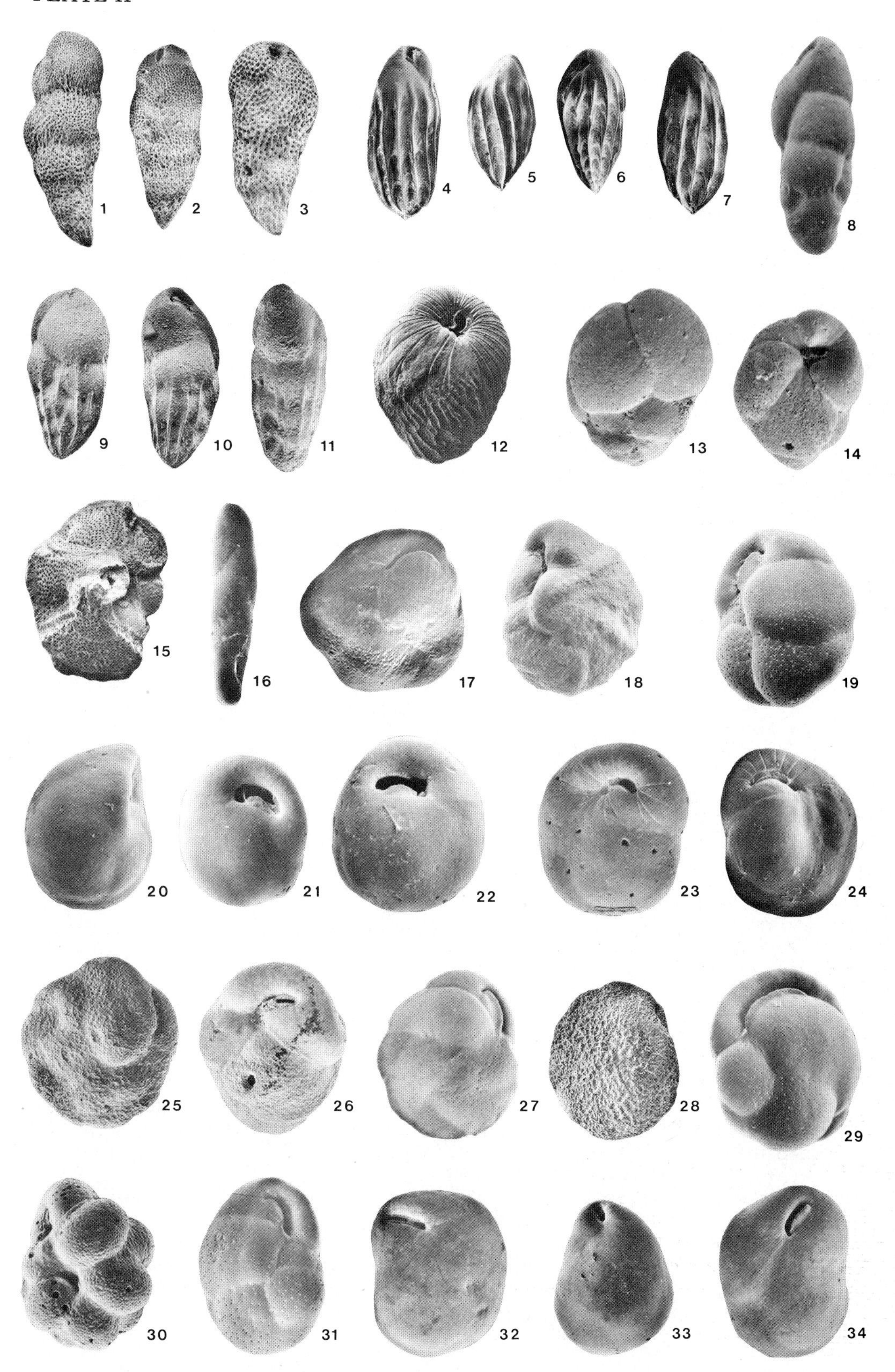
1
2
3
4
5
6
7
8
9
10
11
12
13
14
15
16
17
18
19
20
21
22
23
24
25
26
27
28
29
30
31
32
33
34

Bolivina silvestrina Cushman, 1936, Cushman Lab.Foraminiferal Res., Spec.Publ., 6: p.56, pl.8, f.5. Plio.(254,R).
Bolivina subspinescens Cushman, 1922, U.S.Natl.Mus., Bull., 104(part 3): p.48, pl.7, f.5. Plio.—Quat.(216,R).
Bolivina cf. *B. thalmanni* Renz, 1948, Geol.Soc.Am.,Mem., 32: p.120, pl.12, f.13. Olig.—U.Mio.(214,216,217,253,254, R—F); Plio.(214,216,253,254, R—F)·Quat.(214,253, R). The main difference between *B. thalmanni* and my specimens is that the majority of mine have a tendency to form a quadrilateral test, thus closely relating them to the genus *Bolivinita.*
Bolivina cf. *B. villalverniensis* Martinis, 1954. Riv.Ital.Paleontol. Stratigr. 60(3): p.174, pl.7, f.2—5. Mio.—Plio.(214,253,254,R). My specimens differ from *B. villalverniensis* in being on the average shorter and in having a very rough surface, especially on the aboral half of the test. This roughness is due to the presence of transparent calcareous shell material. The sutures are flush, broad and clearly visible beneath the ornamentation. The shape of the chambers as well as the lobate peripheral margin are identical with those of the primary types. The SEM photographs of my specimens are an example of the taxonomical difficulties which can be created by the use of SEM photographs alone. Since the specimens are coated, the sutures are concealed and thus the peculiar chamber shape is not visible. On the other hand ornamentation is exaggerated. Undoubtedly my specimens are close to *B.* cf. *thalmanni.*
Bolivinopsis cubensis (Cushman and Bermúdez) = *Spiroplectoides cubensis* Cushman and Bermúdez, 1937, Cushman Lab.Foraminiferal Res., Contrib., 13(part 1): p.13, pl.1, f.44,45. Olig.(217,F); L.Mio.(216,R); M.Mio.(214,254,R).
"*Bulava indica*"-Boltovskoy, 1976, Rev. Española Micropaleontol., 8(2): pl.1, f.1—23. M.Mio.(214,216,217,253,254,R); U.Mio.(214,216,253,R—F).
Bulimina inflata Seguenza, 1862, Accad.Gioenia Sci.Nat.Catania, Atti, Ser.2, 18: p.109, pl.1, f.10. M.Mio.—U.Mio.(214,254,R); Plio.—Quat.(254,R).
Bulimina jarvisi Cushman and Parker, 1936, Cushman Lab.Foraminiferal Res., Contrib., 12(part 2): p.39, pl.7, f.1. Olig.(216, 217,R; 214,253,F); L.Mio—M.Mio.(214,253,R).
Bulimina macilenta Cushman and Parker = *Bulimina denticulata* Cushman and Parker, 1936, Cushman Lab.Foraminiferal Res., Contrib., 12(part 2): p.42, pl.7, f.7,8; *Bulimina macilenta* Cushman and Parker, new name, 1939, Cushman Lab.Foraminiferal Res., Contrib., 15(part 4): p.93. Olig.—U.Mio.(214,216,253,R).
Bulimina miolaevis Finlay, 1940, R.Soc.N.Z., Trans.Proc., 69(part 1): p.454, pl.64, f.70,71. Olig.—L.Mio.(254,R—F); M.Mio.(253,254,R).
Bulimina rostrata Brady, 1884, "Challenger" Exped., Rep., Zool., 9: p.408, pl.51, f.14,15. Olig.—U.Mio.(217,R); L.Mio.—Quat.(214,253,R—F; 216,R).
Bulimina semicostata Nuttall, 1930, J.Paleontol., 4(3): p.285, pl.23, f.15,16. Olig.—U.Mio.(214,216,217,253,254,R—F).
Bulimina translucens Parker, 1953, Swed. Deep-Sea Exped., Rep., 7(1): p.33, pl.6, f.30,31. Olig.(254,R); Olig.—Plio.(217,R); Quat.(217,F).
Buliminella carteri Bhatia, 1955, J.Paleontol., 29(4): p.678, pl.66, f.10; tf.4. Olig.—L.Mio.(214,216,253,R); M.Mio.(253,R).
Buliminella sculpturata Keijzer, 1953, Leidse Geol.Meded., 17 (1952): p.276, pl.1, f.20—22. Olig.(253,R); U.Mio.(214,R); Olig.—Plio.(214,R).
Carpenteria balaniformis Gray, 1858, Zool.Soc.London, p.268, tf.1—4. M.Mio.(254,R); Plio.(254,R).
Cassidella bradyi (Cushman) = *Virgulina bradyi* Cushman, 1922, U.S. Natl.Mus., Bull., 104: p.115, pl.24, f.1. M.Mio.(254,R); U.Mio.(217,R); Plio.(216,R); Olig.—Quat.(214,R—F).
Cassidulina carandelli Colom, 1943, R.Soc. Española Hist.Nat., Bol., 41: p.324, pl.23, f.65—67. L.Mio.—M.Mio.(214,253,254,R).
Cassidulina carapitana Hedberg, 1937, J.Paleontol., 11(8): p.680, pl.92, f.6. U.Mio.(214,R).
Cassidulina caudriae Cushman and Stainforth, 1945, Cushman Lab.Foraminiferal Res., Spec.Publ., 14: p.64, pl.12, f.2,3. Olig.—L.Mio.(214,216,R—F); M.Mio.(214,217,R).
Cassidulina crassa d'Orbigny, forma minima Boltovskoy, 1959, Argentina, Serv.Hidr.Nav.,

H.1005: p.100, pl.14, f.12. L.Mio.(216,253,R); M.Mio.—Quat.(214,216,217,R). This foraminifer was found by me in the South Atlantic Ocean and determined at that time temporarily as forma minima of the well-known species *Cassidulina crassa* d'Orbigny (Boltovskoy, loc.cit.). Detailed comments on it were given in the paper cited. Except for size, it is practically identical with typical specimens of *C. crassa*. The latter is usually about 1 mm in length, but forma minima attains only 0.2—0.3 mm. There are no linkage forms between them. Forma minima was later found in other areas, but I have not been able to find a species described which fits this forma. Therefore I continue interpreting it as a forma minima of *C. crassa*.

Cassidulina cuneata Finlay, 1940, R.Soc. N. Z., Trans. Proc., 69(part 1): p.456, pl.63, f.62—66. Olig.—M.Mio.(214,216,217,253,254,F—C); M.Mio(214,216,253,R). The following difference can be mentioned between my specimens and *C. cuneata* as described by Finlay: mine have faint radial grooves, especially well seen in the SEM photographs. However, these grooves are not always well pronounced and in some specimens are barely visible under a common microscope.

Cassidulina elegans Sidebottom, 1910, Quekett Microsc. Club, J., Ser.2,11(67): p.106, pl.4, f.1. Quat.(214,R).

Cassidulina laevigata d'Orbigny, 1826, Ann.Sci.Nat., Sér.1, 7: p.282, no.1, pl.15, f.4,5. U.Mio.(217,R); Plio.—Quat.(214,R).

Cassidulina limbata Cushman and Hughes, 1925, Cushman Lab. Foraminiferal Res., Contrib., 1(part 1): p.12, pl.2, f.2. M.Mio.(254,R); U.Mio.(214,253,R); Plio.(214,R); Quat.(216,R).

Cassidulina minuta Cushman, 1933, Cushman Lab. Foraminiferal Res., Contrib., 9(part 4): p.92, pl.10, f.3. L.Mio.—U.Mio.(214,216,253,R); Plio.(214,216,R); Quat.(253,254, R).

Cassidulina monicana Cushman and Kleinpell, 1934, Cushman Lab.Foraminiferal Res., Contrib., 10(part 1): p.16, pl.3, f.4. L.Mio.(254,R).

Cassidulina oblonga Reuss, 1850, K.Akad.Wiss.Wien,Math.-Nat. Cl., Denkschr., 1: p.376, pl.48, f.5,6. Olig.(253,R); U.Mio.(253,R).

Cassidulina subglobosa horizontalis Cushman and Renz, 1941, Cushman Lab.Foraminiferal Res., Contrib., 17: p.26, pl.4, f.8. Olig.(253,F); L.Mio.—U.Mio.(214,217,253, 254,F—C).

Cassidulina subglobosa producta Chapman and Parr = *Cassidulina subglobosa* Brady, var. *producta* Chapman and Parr, 1937, Australas. Antarct. Exped. Sci. Rep., C, 1: p.82, pl.8, f.12. Plio.(214,R).

Cassidulina subglobosa subglobosa Brady = *Cassidulina subglobosa* Brady, 1884, "Challenger" Exped., Rep., Zool., 9: p.430, pl.54, f.17. Olig. (216,253,R); L.Mio. (214,253,254,R); M.Mio. (214,216,217,253,254,R); U.Mio.—Quat. (214,216,217,253, 254,F.—C).

Cassidulina subtumida Cushman, 1933, Cushman Lab.Foraminiferal Res., Contrib., 9(part 4): p.93, pl.10, f.5. Olig.(217,R); L.Mio.—U.Mio.(214,R—F; 216,R); L.Mio.—Quat.(253,R); Plio.(214,R).

Cassidulinoides tenuis Phleger and Parker, 1951, Geol.Soc.Am., Mem., 46(part 2): p.27, pl.14, f.14—17. U.Mio.(214,216,R).

Chilostomella ovoidea Reuss 1850, K.Akad. Wiss.Wien, Math.-Nat. Cl., Denkschr., 1: p.380, pl.48, f.12. U.Mio.(214,R).

Cibicides aknerianus (d'Orbigny) = *Rotalina akneriana* d'Orbigny, 1846, Foram.Foss. Vienne, p.156, pl.8, f.13—15. L.Mio.—U.Mio.(253,254,R); Plio.(254,R).

Cibicides bradyi (Trauth) = *Truncatulina dutemplei* Brady, 1884, "Challenger" Exped., Rep., Zool., 9: p.665, pl.95, f.5 = *Truncatulina bradyi* Trauth, 1918, K.Akad.Wiss. Wien, Math.-Nat. Kl., Denkschr., 95: p.235. L.Mio.—M.Mio.(253,R); U.Mio.(253, 254,R); Olig.—Quat.(214,216,217,R—F).

Cibicides kullenbergi Parker, 1953, In: Phleger, Parker and Peirson, 1953, *Swed. Deep-Sea* Exped., Rep., 1: p.49, pl.11, f.7,8. Olig.—U.Mio.(217,F); Olig.—Quat.(214,216,253, 254,F —C).

PLATE III

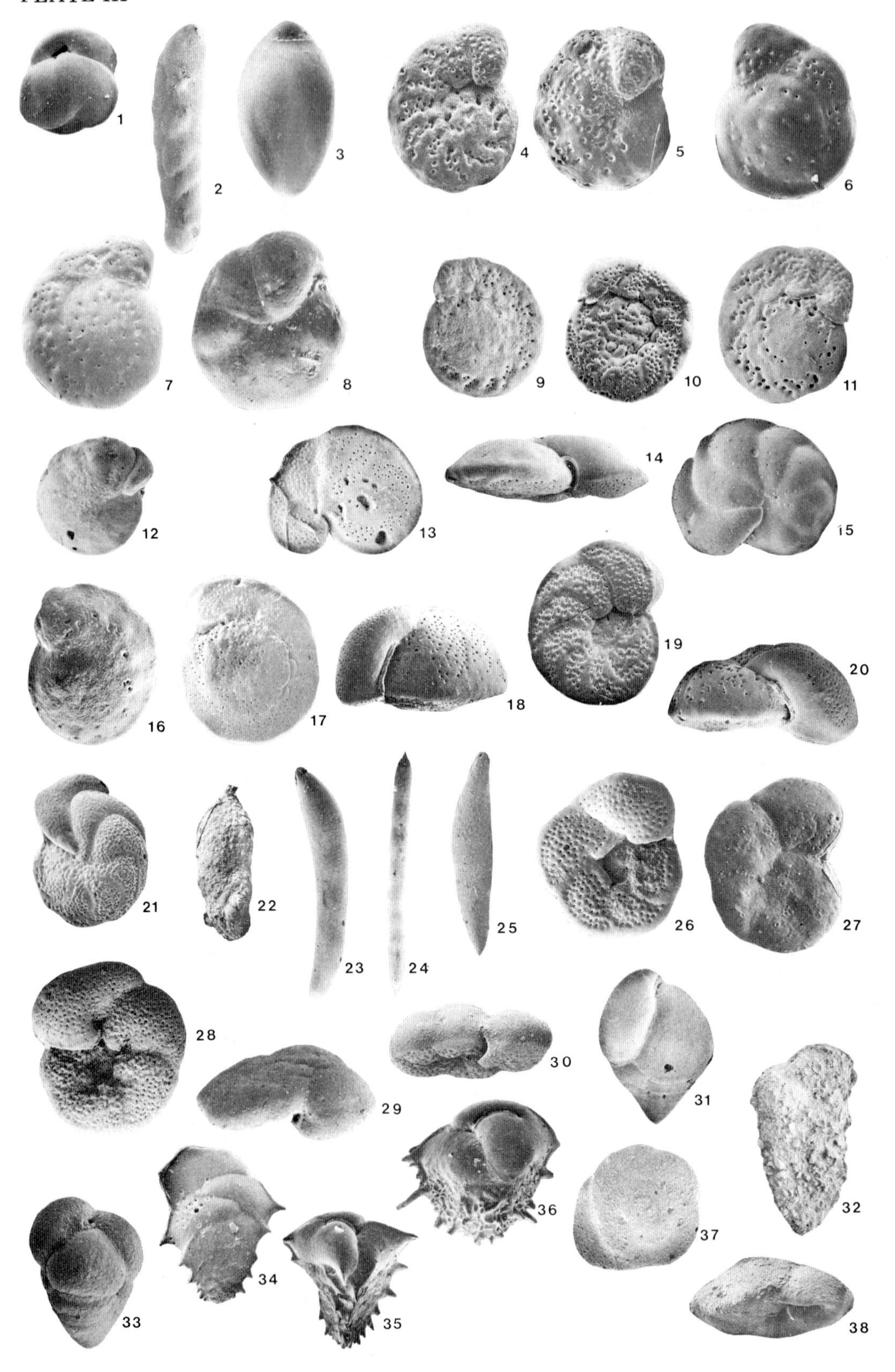
1
2
3
4
5
6
7
8
9
10
11
12
13
14
15
16
17
18
19
20
21
22
23
24
25
26
27
28
29
30
31
32
33
34
35
36
37
38

?*Cibicides lucidus* (Reuss) = *Truncatulina lucida* Reuss, 1866, K.Akad.Wiss., Wien, Math.-Naturw. Cl., Denkschr., 25(1): p.160, pl.4, F.15. Olig.—L.Mio.(254,F—C); M.Mio.(254, R); U.Mio.(217,254,R); Plio.—Quat.(254,R).
Reuss (loc. cit.) writes that the test is coarsely perforate. However, according to his figures, the involute side of his specimens does not have pores. This is the case with my specimens also. Another discrepancy between the description and the figures given by Reuss is in the number of chambers in the last coil. Again my specimens are closer to the figure which has 9 chambers. Thus, the present identification is a tentative one.

Cibicides notocenicus Dorreen = *Cibicides perforatus* (Karrer) var. *notocenicus* Dorreen, 1948, J.Paleontol., 22: p.299, pl.41, f.4. Olig.—L.Mio.(253,254,F—C); M.Mio.(214,R); U.Mio.(214,254,I).

Cibicides wuellerstorfi (Schwager) = *Anomalina wuellerstorfi* Schwager, 1866, "Novara" Exped., Geol., 2: p.258, pl.7, f.105—107. L.Mio.(254,R); M.Mio.—Quat.(214,216,217, 253,254,F—C).
This species is extremely variable, expecially as far as porosity and character of the sutures is concerned.

Cylindroclavulina bradyi (Cushman) = *Clavulina bradyi* Cushman, 1911, U.S.Natl.Mus., Bull., 71(part 2): p.73, f.118,119. L.Mio.—M.Mio. (254,R).

Dentalina communis d'Orbigny = *Nodosaria* (*Dentalina*) *communis* d'Orbigny, 1826, Ann.Sci.Nat., Sér.1, 7: p.254. Olig.—Plio.(214,216,217,253,254,R); Quat.(253,254, R).

Dentalina intorta (Dervieux) = *Nodosaria intorta* Dervieux, 1894, Soc.Geol.Ital., Boll., 12(4): p.610, pl.5, f.32—34. Olig.(214,R); L.Mio.(214, 254,R); M.Mio.(214,R); U.Mio. (253,R).

Dentalina reussi Neugeboren, 1856, K.Akad, Wiss. Math-Naturw. Cl., Denkschr., 12 (Abt. 2): p.85, pl.3, f.6,7,17. Olig.—U.Mio.(214,216,217,253,254,R).

Discorbis subvilardeboanus (Rhezak) = *Discorbina sub-vilardeboana* Rhezak, 1888, Naturhist. Mus., Ann., 3: p.263, pl.11, f.6. Olig.—L.Mio.(214,C; 216,R).

PLATE III

1. *Cassidulina subtumida* Cushman; × 50; site 253.
2. *Cassidulinoides tenuis* Phleger and Parker; × 75; site 214.
3. *Chilostomella ovoidea* Reuss; × 45; site 214.
4, 5. *Cibicides aknerianus* (d'Orbigny); × 25; site 254.
6—8. *Cibicides bradyi* (Trauth); × 50; site 216.
9—12. *Cibicides kullenbergi* Parker; × 25 (figs.9,10), × 40 (figs.11,12); site 253 (figs.9,10), site 254 (figs.11,12).
13—15. ?*Cibicides lucidus* (Reuss); × 35; site 254.
16—18. *Cibicides notocenicus* Dorreen; × 35; site 254.
19—21. *Cibicides wuellerstorfi* (Schwager); × 25 (figs.19,20), × 20 (fig.21); site 254 (figs.19,20), site 253 (fig.21).
22. *Cylindroclavulina bradyi* (Cushman); × 15; site 254.
23. *Dentalina communis* d'Orbigny; × 35; site 253.
24. *Dentalina reussi* Neugeboren; × 15; site 216.
25. *Dentalina intorta* (Dervieux); × 30; site 217.
26—30. *Discorbis subvilardeboanus* (Rhezak); × 40 (figs.26,30), × 30 (figs.27—29); site 214 (figs.26,29), site 216 (figs.27,28,30).
31. *Dorothia brevis* Cushman and Stainforth; × 20; site 253.
32. *Dorothia scabra* (Brady); × 65; site 254.
33. *Eggerella bradyi* (Cushman); × 25; site 253.
34, 35. *Ehrenbergina carinata* Eade; × 35; site 253.
36. *Ehrenbergina hystrix* Brady; × 35; site 214.
37, 38. *Epistominella exigua* (Brady); × 70; site 253.

Dorothia brevis Cushman and Stainforth, 1945, Cushman Lab.Foraminiferal Res., Spec. Publ., 14: p.18, pl.2, f.5. Olig.(214,216,217,R); L.Mio.(214,253,R); M.Mio.(214,216, 217,253,254,R); U.Mio.—Plio.(253,R).

Dorothia scabra (Brady) = *Gaudryina scabra* Brady, 1884, "Challenger" Exped., Rep., Zool., 9: p.381, pl.46, f.14—16. L.Mio.(254,R); Plio.—Quat.(254,R).

Eggerella bradyi (Cushman) = *Verneuilina bradyi* Cushman, 1911, U.S.Natl.Mus., Bull., 71(part 2): p.54, tf.87. Olig.—Plio.(214,216,217,253,254,R—F); Quat.(216,253,254, R).

Ehrenbergina carinata Eade, 1967, N.Z.J.Mar.Freshwater Res., 1(4); p.448,450, tf.8,9. M.Mio.—U.Mio.(254,R—F); Plio.(214,217,254,R; 216,C); Quat.(214,216,253,254,R).

Ehrenbergina hystrix Brady, 1884, "Challenger" Exped., Rep., Zool., 9: p.435, pl.55, f.8—11. L.Mio.—M.Mio.(254,R); U.Mio.(216,217,254,R; 253,F); Quat.(216,R).

Epistominella exigua (Brady) = *Pulvinulina exigua* Brady, 1884, "Challenger" Exped., Rep., Zool., 9: p.696, pl.103, f.13,14. Olig.—M.Mio.(216,217,F—C; 214,253,R); U.Mio.—Plio.(214,216,217,F—C); Quat.(214,216,217,F—C; 253,R).

Eponides bradyi Earland = *Truncatulina pygmaea* Hantken, in Brady, 1884, "Challenger" Exped., Rep., Zool., 9: p.666, pl.95, f.9,10 = *Eponides bradyi* Earland, 1934, "Discovery" Exped., Rep., 10: p.187, pl.8, f.36—38. U.Mio.(253,R); Olig.—Quat.(214, 216,217,R—F).

Eponides polius Phleger and Parker, 1951, Geol.Soc.Am., Mem., 46(part 2): p.21, pl.11, f.1,2. Olig.—L.Mio.(253,R); M.Mio.(254,R); Olig.—U.Mio.(214,216,217,R—F); Plio.—Quat.(214,216,R—F; 217,C).

Eponides weddellensis Earland, 1936, "Discovery" Rep., 13: p.57, pl.1, f.65—67. M.Mio. (214,216,217,R); U.Mio.—Quat.(217,C); Plio.(214,R); Quat.(216,R).

PLATE IV

1—3. *Eponides bradyi* Earland; × 75 (fig.1), × 50 (figs.2,3); site 216 (fig.1), site 253 (figs.2,3).
4, 5. *Eponides polius* Phleger and Parker; × 85 (fig.4), × 75 (fig.5), site 253.
6—8. *Francesita advena* (Cushman); × 65 (fig.6), × 50 (fig.7), × 75 (fig.8, apertural view); site 214 (fig.6), site 253 (figs.7,8).
9. *Frondicularia advena* Cushman; × 30; site 253.
10, 11. *Gaudryina trinitatensis* Nutall; × 20 (fig.10), × 30 (fig.11, apertural view); site 254.
12, 13. *Gavelinopsis lobatulus* (Parr); × 100; site 254.
14, 15. *Gyroidina lamarckiana* (d'Orbigny), s.l.; × 40; site 253.
16, 17. *Gyroidina orbicularis* d'Orbigny; × 50; site 214.
18, 19. *Gyroidina soldanii* d'Orbigny; × 35; site 253 (fig.19), site 214 (fig.18).
20. *Favocassidulina favus* (Brady); × 25; site 253.
21. *Gyroidina umbonata* (Silvestri); × 85; site 219.
22—26. *Heronallenia* sp. "A"; × 45 (figs.22—24), × 75 (fig.25), × 100 (fig.26); site 253 (figs.22,23), site 216 (fig.24,26), site 217 (fig.25).
27. *Karreriella alticamera* Cushman and Stainforth; × 35; site 214.
28, 29. *Karreriella bradyi* (Cushman); × 20; site 253.
30, 31. *Kyphopyxa* sp. "A"; × 50; (fig.30), × 65 (fig.31); site 253.
32. *Laticarinina pauperata* (Parker and Jones); × 25; site 253.
33. *Martinottiella antarctica* (Parr); × 15; site 219.
34, 35. *Martinottiella scabra* (Cushman); × 15 (fig.34), × 10 (fig.35); site 253 (fig.34) site 216 (fig.35).
36. *Nodosarella pacifica* Cushman; × 25; site 216.
37. ?*Nodosaria sulcata* Nilson; × 50; site 253.
38. *Nodosaria vertebralis* (Batsch); × 10; site 253.
39. *Nodosaria fusiformis* Silvestri; × 50; site 253.

PLATE IV

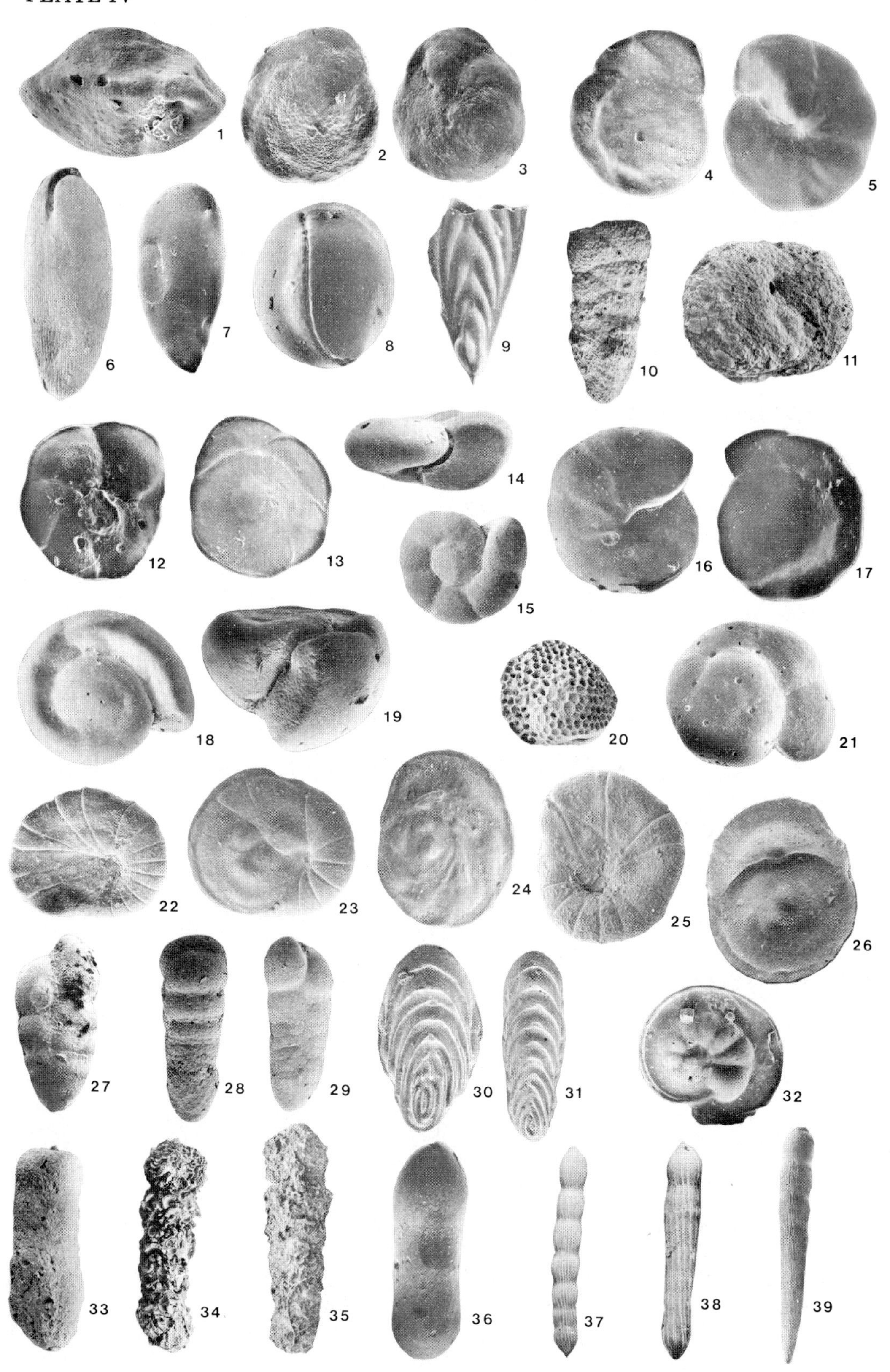
1
2
3
4
5
6
7
8
9
10
11
12
13
14
15
16
17
18
19
20
21
22
23
24
25
26
27
28
29
30
31
32
33
34
35
36
37
38
39

Favocassidulina favus (Brady) = *Pulvinulina favus* Brady, 1884, "Challenger" Exped., Rep., Zool., 9: p.701, pl.104, f.12—16. M.Mio.—U.Mio.(214,216,217,R).
Francesita advena (Cushman) = *Virgulina* (?) *advena* Cushman, 1922, U.S.Natl.Mus., Bull., 104(part 3): p.120, pl.25, f.1—3. M.Mio.(216,217,R); U.Mio.(214,216,217,253, R); Quat.(217,253,R).
Frondicularia advena Cushman, 1923, U.S.Natl.Mus., Bull., 104(part 4): p.141, pl.20, f.1,2. M.Mio.(253,R).
Gaudryina trinitatensis Nuttall, 1928, Geol.Soc.London, Q.J., 84: p.76, pl.3, f.15,16. Olig.—L.Mio.(254,R).
Gavelinopsis lobatulus (Parr) = *Discorbis lobatulus* Parr, 1950, B.A.N.Z.Antarct.Res. Exped., Rep., Ser.B,5(part 6): p.354, pl.13, f.23—25. Plio.(254,R).
Gyroidina lamarckiana (d'Orbigny), s.l. = *Rotalina lamarckiana* d'Orbigny, 1839, in Barker, Webb and Berthelot, Hist.Nat. Iles Canaries, 2(2): p.131, pl.2, f.13—15. Olig.—U.Mio.(214,216,217,F; 253,R); M.Mio.—U.Mio.(254,R); Plio.(214,253,254,R; 216,217,F); Quat.(214,253,R;216,217,C).
Gyroidina orbicularis d'Orbigny, 1826, Ann.Sci.Nat., Sér.1, 7: p.278, Mod. no.13. Plio.—Quat.(214,R).
Gyroidina soldanii d'Orbigny, 1826, Ann.Sci.Nat., Sér.1, 7: p.276, no.5, Mod. no.36. Olig.—U.Mio.(214,216,217,253,F; 254,R); Plio.(214,216,254,R; 253,F); Quat.(214, 216,217,253,254,R).
Gyroidina umbonata (Silvestri) = *Rotalia soldanii* d'Orbigny var. *umbonata* Silvestri, 1898, Accad.Pont. Nuovi Lincei, Mem., 15: p.329, pl.6, f.14. Olig.—L.Mio.(216,217, R); M.Mio.—U.Mio.(214,216,217,254,R); Plio.—Quat.(214,216,R; 217,C).
Heronallenia sp. "A" Olig.—U.Mio.(214,216,217,253,254,R); Plio.(216,217,254,R). There are few described species of *Heronallenia* and none of them fit my specimens. Mine resemble *H. cubana* Palmer and Bermúdez, but differ from it mainly in the character of both the sutures and the ventral side ornamentation. *H. cubana* has broadly curved depressed sutures. The sutures of my specimens are flush or raised and

PLATE V

1,2. *Nonion affine* (Reuss); × 50; site 214.
3, 4. *Nonion pompilioides* (Fichtel and Moll); × 50; site 253.
5, 6. *Oridorsalis umbonatus* (Reuss); × 45; site 217 (fig.5), site 253 (fig.6).
7—11. *Orthomorphina antillea* (Cushman); × 35 (fig.7), × 45 (fig.8), × 60 (figs.9—11); site 253 (figs.7—9), site 254 (figs.10,11).
12—15. *Orthomorphina* aff. *O. antillea* (Cushman); × 55 (figs.12—14), × 65 (fig.15), × 15 (fig.13); site 253.
16, 17. *Orthomorphina challengeriana* (Thalmann); × 45; site 253.
18. *Orthomorphina columnaris* (Franke); × 40, site 254.
19. *Orthomorphina glandigena* (Schwager); × 30; site 216.
20—22. *Orthomorphina himerensis* (de Amicis); × 30 (fig.20), × 35 (fig.21), × 40 (fig.22); site 253.
23, 24. *Orthomorphina perversa* (Schwager); × 50; site 217.
25. *Orthomorphina modesta* (Bermúdez); × 40; site 214.
26. *Orthomorphina richardsi* (McLean); × 35; site 253.
27, 28. *Orthomorphina scalaris* (Batsch); × 35; (fig.27), × 50 (fig.28); site 254.
29—34. *Osangularia culter* (Parker and Jones); × 25 (figs.29,30,32,33), × 30 (fig.31), × 35 (fig.34); site 253.
35—38. *Planulina marialana gigas* Keijzer; × 20; (figs.35,36), × 25 (figs.37,38); site 254 (figs.35,36), site 253 (figs.37,38).
39—41. *Pleurostomella acuminata* Cushman; × 30 (figs.39,41), × 45 (fig.40); site 253.
42. *Pleurostomella acuta* Hantken; × 35; site 253.
43, 44. *Pleurostomella alternans* Schwager; × 20 (fig.43), × 30 (fig.44); site 253.
45. *Pleurostomella bierigi* Palmer and Bermúdez; × 30; site 253.

PLATE V

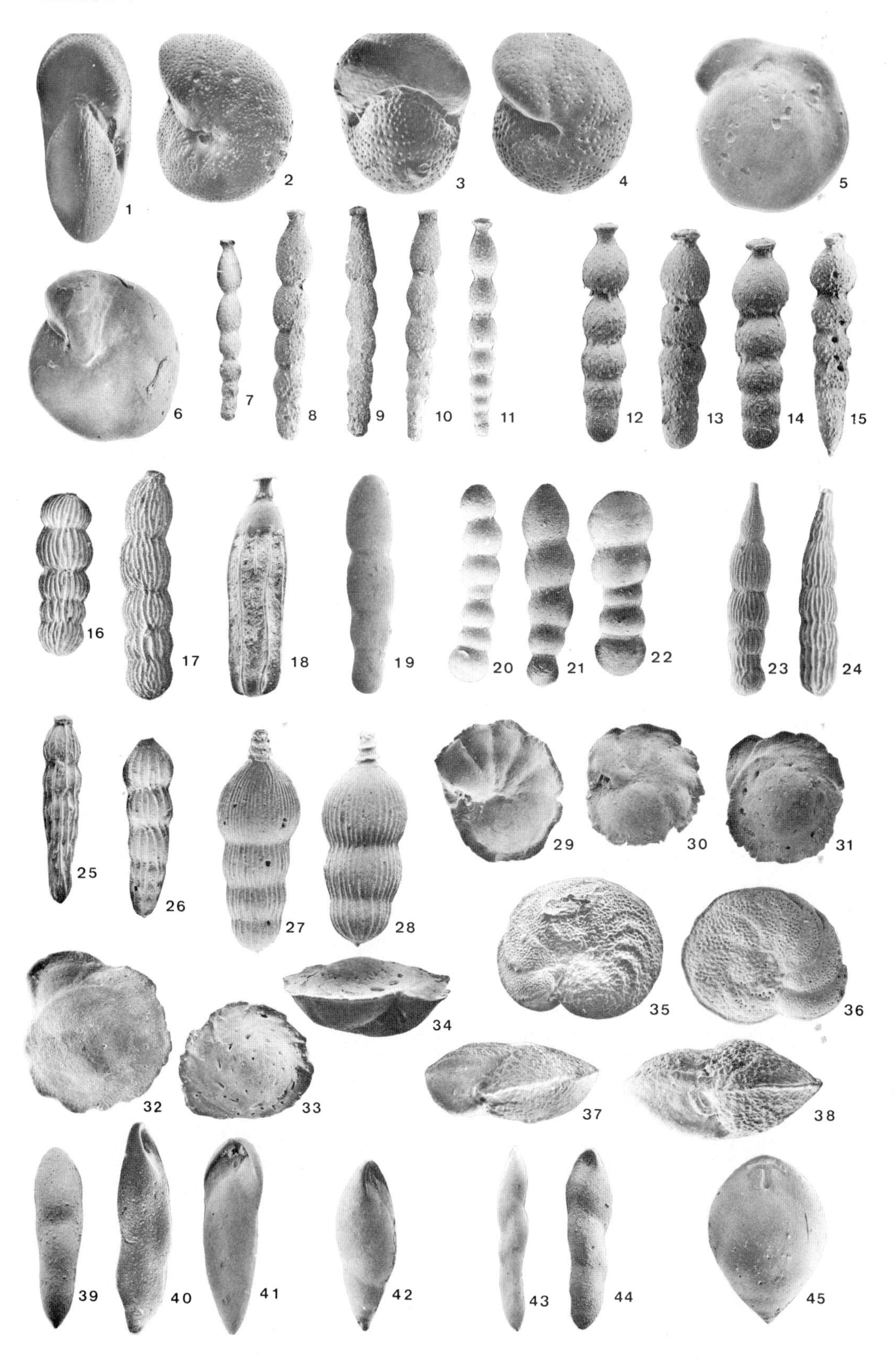

1
2
3
4
5
6
7
8
9
10
11
12
13
14
15
16
17
18
19
20
21
22
23
24
25
26
27
28
29
30
31
32
33
34
35
36
37
38
39
40
41
42
43
44
45

very irregular in width. The ventral surface in *H. cubana* is covered with delicate costae while in my specimens it is radially finely furrowed. It is quite probable that the specimens discussed should be assigned to a new species. For the time being, however, I prefer to leave them in *nomenclatura aperta*.

Karreriella alticamera Cushman and Stainforth, 1945, Cushman Lab.Foraminiferal Res., Spec. Publ., 14: p.19, pl.2, f.10. Olig.(214,253,254,R); L.Mio.—M.Mio.(214,R).

Karreriella bradyi (Cushman) = *Gaudryina bradyi* Cushman, 1911, U.S.Natl.Mus., Bull., 71(part 2): p.67, tf.107. Olig.—U.Mio.(214,F; 216,217,253,254,R); Plio.(214,253, R; 254,F); Quat.(214,216,253,R; 254,F).

Kyphopyxa sp. "A" Olig.(253,R); L.Mio.(214,R); M.Mio.(214,253,254,R).
A few specimens which belong to this genus were found. This is an interesting find, as up to now (as far as I know) the genus *Kyphopyxa* has not been known in deposits younger than Cretaceous. The specimens found apparently represent a new species; it is not described here because of the small number of specimens available.

Laticarinina pauperata (Parker and Jones) = *Pulvinulina repanda* var. *menardii* subvar. *pauperata* Parker and Jones, 1865, R.Soc.London, Philos.Trans., 155: p.395, pl.16, f.50,51. Olig.—L.Mio.(253,254,R); M.Mio.—Plio.(217,R); M.Mio.—Quat.(214,216, 253,254,R).

Martinottiella antarctica (Parr) = *Schenckiella antarctica* Parr, 1950, B.A.N.Z. Antarct. Res. Exped., Rep., Ser. B, 5(part 6): p.284, pl.5, f.27. U.Mio.(214,217,R).

Martinottiella scabra (Cushman) = *Pseudoclavulina scabra* Cushman, 1936, Cushman Lab.Foraminiferal Res., Spec.Publ., 6: p.20, pl.3, f.11. Olig.—Plio.(214,216,253,254, R).

Nodosarella pacifica Cushman, 1931 Cushman Lab.Foraminiferal Res., Contrib., 7(part 2): p.31, pl.4, f.12,13. Olig.(253,R); L.Mio.(214,216,254,R); M.Mio.(214,216,217, 254,R).

Nodosaria fusiformis Silvestri, 1872, Accad. Gioenia Sci.Nat. Catania, Atti, Ser.3, 7: p.99, f.34. In: Costa, 1855, Foram.Terz. Messina. Olig.(214,216,253,R); L.Mio.(254, R); M.Mio.(217,253,254,R); U.Mio.(214,217,R); Plio.(214,R).

?*Nodosaria sulcata* Nilson, 1826, in Brotzen, 1937, Geol.Foren. Stockholm Förhandl., 59(1): pl.2, f.8—16. Olig.—U.Mio.(214,216,253,254,R).
Several specimens were encountered in the population of *N. vertebralis* which differ in having better pronounced sutures, more numerous ribs and somewhat more enlarged chambers. These specimens are tentatively interpreted as *N. sulcata*.

Nodosaria vertebralis (Batsch) = *Nautilus* (*Orthoceras*) *vertebralis* Batsch, 1791, Conchyl. Seesandes, p.3, no.6, pl.2, f.6. Olig.—U.Mio.(214,216,217,253,254,R); Plio.(253,254, R).

Nonion affine (Reuss) = *Nonionina affinis* Reuss, 1851, Dtsch. Geol. Ges., Z., 3: p.72, pl.5, f.32. L.Mio.(214,254,R); M.Mio.—U.Mio.(214,216,217,253,R—F); Plio.(214,217, 253,254,R; 216,F); Quat.(217,R).

Nonion pompilioides (Fichtel and Moll) = *Nautilus pompilioides* Fichtel and Moll, 1798, Test. Microsc., p.31, pl.2, f.a—c. U.Mio(217,F); Plio.(253,R).

Oridorsalis umbonatus (Reuss) = *Rotalina umbonata* Reuss, 1851, Dtsch. Geol.Ges., Z., 3: p.75, pl.5, f.35. Olig.—Quat.(214,216,217,253,F—C; 254,R).
There is some confusion with respect to the secondary apertures close to the umbilicus which were not mentioned by Andersen in his description of this genus, but were found by Parker (1964). Todd (1965) also did not find them. The secondary apertures were observed in a few tests in my specimens. In some samples specimens were also found which exhibited sinuously curved sutures on the ventral side. They were described by Silvestri as var. *stellata*.

Orthomorphina antillea (Cushman) = *Nodosaria antillea* Cushman, 1923, U.S.Natl.Mus., Bull., 104(part 4): p.91, pl.14, f.9. M.Mio.—Plio.(214, 253,254,R).
I am not very happy with this identification because my specimens have chambers which are not really angled near the base; but the figure and description of *O. antillea* given by Cushman best fit my specimens.

Orthomorphina aff. *O. antillea* (Cushman) = *Nodosaria antillea* Cushman, 1923, U.S.Natl. Mus., Bull., 104(part 4): p.91, pl.14, f.9. Olig.—U.Mio.(214,216,217,253,254,R—C); Plio.(214,253,254,R).
This species is more widely distributed in the Quaternary and Neogene sediments of the Indian Ocean than *O. antillea.* Undoubtedly it is close to *O. antillea,* differing mainly in having less elongated chambers which are not at all angular near the base.

Orthomorphina challengeriana (Thalmann) =*Nodosaria perversa* Brady, 1884, [not Schwager], "Challenger" Exped., Rep., Zool., 9: p.512, pl.64, f.25—27; *Nodogenerina challengeriana* Thalmann, 1937, Eclogae Geol. Helv., 30: p.341, Olig.(217,254,R); L.Mio.—U.Mio. (214,216,217,253,254,R); Plio.(217,253,254,R).

Orthomorphina columnaris (Franke) = *Nodosaria columnaris* Franke, 1936, Preuss.Geol. Landesanst., Abh., N.F., Heft 169: p.48, pl.4, f.19. L.Mio.—M.Mio.(253,254,R).
The specimens found fit this species well and I ascribed them to it for this reason in spite of their quite different ages (Franke described *O. columnaris* in Jurassic sediments).

Orthomorphina glandigena (Schwager) = *Nodosaria glandigena* Schwager, 1866, "Novara" Exped., Geol., 2: p.219, pl.4, f.46. L.Mio.—U.Mio.(216,253,254,R).

Orthomorphina himerensis (de Amicis) = *Nodosaria himerensis* de Amicis, 1895, Nat.Sci., 14(4—5): p.70, pl.1, f.1. L.Mio.(214,253,254,R); L.Mio.—Plio.(214,217,253,254,R).

Orthomorphina modesta (Bermúdez) = *Ellipsonodosaria modesta* Bermúdez, 1937, Soc. Cubana Hist.Nat., Mem., 11: p.238, pl.20, f.3. Olig. (217,253,254,R); L.Mio.—M.Mio. (214,216,217,253,254,R); U.Mio.(214,253,254,R).

Orthomorphina perversa (Schwager) = *Nodosaria perversa* Schwager, 1866, "Novara" Exped., Geol., 2: p.212, pl.5, f.29. U.Mio.(214,217,R).

Orthomorphina richardsi (McLean) = *Nodosaria richardsi* McLean, 1952, Acad. Nat. Sci. Philadelphia, Not. Nat., No.242: p.7, pl.2, f.13—16. Olig.(253,254,R); L.Mio. (254,R); M.Mio.(214,254,R); U.Mio.(216,R).

Orthomorphina scalaris (Batsch) = *Nautilus* (*Orthoceras*) *scalaris* Batsch, 1791, Conchyl. Seesandes, No.4: pl.2, f.4. L.Mio.—M.Mio.(254,F).

Osangularia culter (Parker and Jones) = *Planorbulina culter* Parker and Jones, 1865, R.Soc. London, Philos. Trans., 155: p.421, pl.19, f.1. Olig.(214,217,R; 253,F); L.Mio.(214, 253,254,R—F); M.Mio.—U.Mio.(214,216,217,253,254,R—F); Plio.(214,253,R—F); Quat.(214,216,253,R).
There is some confusion concerning the aperture of this species. According to Loeblich and Tappan (1964), the genus *Osangularia* has an aperture in the form of an opening at the base of the final chamber. This opening extends at an oblique angle up the aperture face. Todd (1965) described the aperture of this species as being in the form of an elongate opening situated at right angles to the base of the apertural face. None of these authors noticed the presence of supplementary apertures on the evolute side of the species in question. They were also not mentioned by the authors of this species. However, those supplementary apertures are well observed in many specimens (Pl. V, f.31,33,34). They can hardly be orifices made by predators or parasites because of their form and very regular position. In some specimens they are not present (pl. V, f.32).

Planulina marialana gigas Keijzer, 1945, Utrecht Univ., Geogr. Geol. Meded. Physiogr.-Geol. Reeks. Ser.2, No.6: p.206, pl.5, f.77. Olig. (214,217,R; 216,253,F); L.Mio.(214, 216,217,253,R—F); M.Mio.(214,216,217,253,254,R).
My specimens differ from the primary types in usually having a somewhat smaller number of chambers in the last coil. The close relations between this species and *Cibicides wuellerstorfi* are described in this paper in the chapter on evolutionary trends.

Pleurostomella acuminata Cushman, 1922, U.S.Natl.Mus., Bull., 104(part 3): p.50, pl.19, f.6. Olig.(214,216,217,253,254,R); L.Mio.(216,R; 217,C); M.Mio.(214,216,217,F; 253, 254,R); U.Mio.—Plio.(214,216,217,R—F).
With the representatives of the genus *Pleurostomella* I had great taxonomical difficulties. The genus *Pleurostomella* was continuous but never abundant in the sequences

studied at all the sites. It was easy to separate the specimens into the 10 species listed below, because almost all of them fit well the descriptions and figures of these taxa. But whether all these species have zoological value is questionable. I strongly suspect that some of them are synonyms. However, I do not lump them together because several species were represented by a limited number of specimens and as a result the whole chain of transitional forms cannot be reconstructed. In order to do this a thorough taxonomical study of all the original material should be made and this is not possible now.

Pleurostomella acuta Hantken, 1875, K.Ungar. Geol. Anst., Mitt., 4(1): p.44, pl.13, f.18. Olig.(216,217,F); L.Mio.—Plio.(214,216,217,253,254,R).

Pleurostomella alternans Schwager, 1866, "Novara" Exped., Geol., 2: 238, pl.6, f.79 [not f.80]. Olig.(216,R); L.Mio.—M.Mio.(216,217,R—F); Olig.—Plio.(214,253,254, R).

Pleurostomella bierigi Palmer and Bermúdez, 1936, Soc. Cubana Hist. Nat., Mem., 10: p.294, pl.17, f.7,8. Olig.—L.Mio.(214,216,R); Olig.—U.Mio.(253,254,R).

Pleurostomella bolivinoides Schubert, 1911, Geol. Reichsanst. Wien, Abh., 20(4): p.57, tf.4. Olig.(214,253,R); L.Mio.(214,216,254,R); M.Mio.—U.Mio.(214,253,254,R); Plio.(253,R).

Pleurostomella dominicana Bermúdez = *Pleurostomella schuberti* Cushman and Harris var. *dominicana* Bermúdez, 1949, Cushman Lab. Foraminiferal Res., Spec. Publ., 25: p.230, pl.14, f.69,70. Olig.(214,216,253,R); L.Mio.—Plio.(214,216,217,253,254,R).

Pleurostomella obtusa Berthelin, 1880, Soc.Géol. Fr., Mém., Sér.3. 1(5): p.29, pl.1, f.9. Olig.—U.Mio.(214,253,R; 216,217,F); Plio.(214,217,R; 216,F).

PLATE VI

1.	*Pleurostomella bolivinoides* Schubert; × 45; site 214.
2, 3.	*Pleurostomella dominicana* Bermúdez; × 65; (fig.2), × 75 (fig.3); site 214 (fig.2), site 254 (fig.3).
4, 5.	*Pleurostomella obtusa* Berthelin; × 30; site 253.
6.	*Pleurostomella* cf. *P. praegerontica* Cushman and Stainforth; × 40; site 216.
7.	*Pleurostomella* cf. *P. rimosa* Cushman and Bermúdez; × 65; site 254.
8.	*Pleurostomella* sp. 2 of Parker, 1964; × 25; site 214.
9.	*Pseudonodosaria torrida* (Cushman); × 75; site 216.
10, 11.	*Pseudononion japonicum* Asano; × 75; site 216.
12.	*Pullenia bulloides* (d'Orbigny); × 75; site 253.
13, 14.	*Pullenia multilobata* Chapman; × 45; site 253.
15—18.	*Pullenia osloensis* Feyling-Hanssen; × 110; site 216 (figs.15,16), site 217 (fig.17), site 254 (fig.18).
19.	*Pullenia salisburyi* Stewart and Stewart; × 50; site 253.
20.	*Pullenia quadriloba* Reuss; × 60; site 216.
21, 22.	*Pullenia subcarinata subcarinata* (d'Orbigny); × 80; site 253.
23, 24.	*Pullenia subcarinata quinqueloba* (Reuss); × 70 (fig.23), × 40 (fig.24); site 253.
25.	*Pyrgo depressa* (d'Orbigny); × 20; site 253.
26.	*Pyrgo murrhina* (Schwager); × 30; site 253.
27, 28.	*Pyrgo* aff. *P. nasuta* Cushman; × 35; site 253.
29.	*Pyrgo lucernula* (Schwager); × 15; site 253.
30.	*Pyrulina fusiformis* (Roemer); × 65; site 254.
31.	*Pyrgo serrata* (Bailey); × 30; site 217.
32, 33.	*Quinqueloculina venusta* Karrer; × 35 (fig.32), × 50 (fig.33, apertural view); site 253.
34—36.	*Quinqueloculina* cf. *Q. pygmaea* Reuss; × 100 (figs.34,35), × 85 (fig.36); site 253.
37.	*Quinqueloculina weaveri* Rau; × 30; site 253.
38, 39.	*Rectuvigerina royoi* Bermúdez and Fuenmayor; × 25 (fig.38), × 20 (fig.39); site 253.

PLATE VI

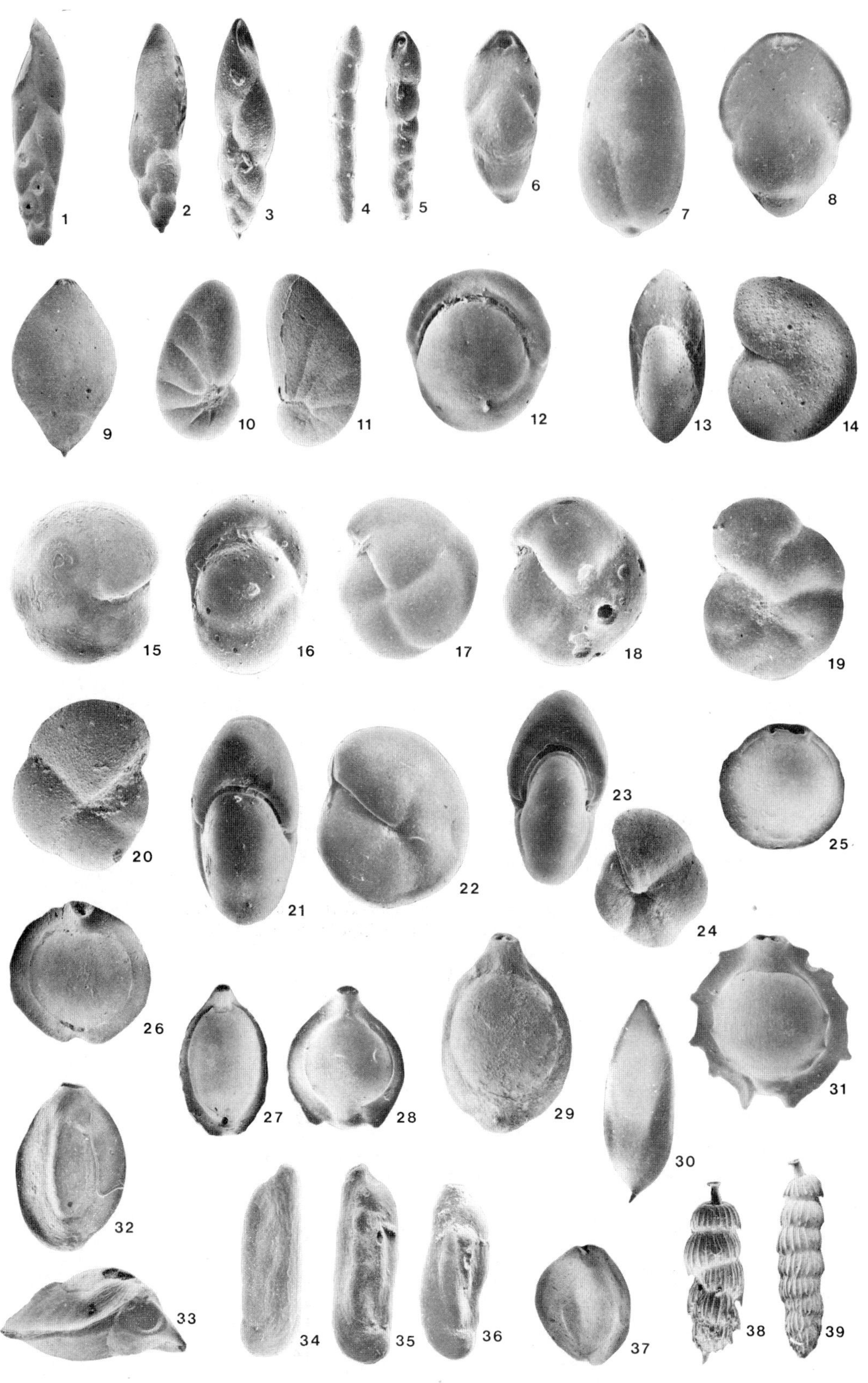
1
2
3
4
5
6
7
8
9
10
11
12
13
14
15
16
17
18
19
20
21
22
23
24
25
26
27
28
29
30
31
32
33
34
35
36
37
38
39

Pleurostomella cf. *P. praegerontica* Cushman and Stainforth, 1945, Cushman Lab. Foraminiferal Res., Spec.Publ., 14: p.52, pl.8, f.13,14. Olig.—L.Mio.(214,217,253,254R; 216,F); M.Mio.(214,253,254,R).
The typical *P. praegerontica* has more inflated chambers than my specimens.
Pleurostomella cf. *P. rimosa* Cushman and Bermúdez, 1937, Cushman Lab. Foraminiferal Res., Contrib., 13(part 1): p.17, pl.1, f.62,63. Plio.(254,R).
Pleurostomella "sp. 2" Parker, 1964, J.Paleontol., 38(4): p.627, pl.99, f.15. Olig.—M.Mio.(214,R).
One specimen of this species was found by Parker (loc. cit.) in that part of the Miocene sediments of the Mohole Drilling which she determined to be late Middle Miocene. In the present material 10 specimens more were found which undoubtedly belong to the same species. For the time being, in view of insufficient number of specimens, I prefer to use Parker's provisory designation. They were found in the Oligocene—M.Miocene sequence of the Ninetyeast Ridge and also in L.Mio.—M.Mio. sediments at site 251 (Boltovskoy, in prep).
Pseudonodosaria torrida (Cushman) = *Nodosaria (Glandulina) laevigata*, var. *torrida* Cushman, 1923, U.S.Natl.Mus., Bull., 104(part 4): p.65, pl.12, f.10. Olig.—Plio.(254,R).
Pseudononion japonicum Asano, 1936, Geol.Soc. Jpn, J., 43(512): p.347, tf.A—C. L.Mio.—Plio.(216,R); Quat.(216,217,R).
The specimens encountered are rather poor representatives of this species; they are small-sized and have fewer chambers.
Pullenia bulloides (d'Orbigny) = *Nonionina bulloides* d'Orbigny, 1846, Foram.Foss. Vienne, p.107, pl.5, f.9,10. Olig.—U.Mio.(216,217,R); Olig.—Plio.(214,253,254,R); Quat.(214,253,R).
Pullenia multilobata Chapman, 1900, Calif.Acad.Sci., Proc., Ser.3, 1(8): p.253, pl.30, f.7. Olig.(214,217,R); L.Mio.(214,253,R; 217,C); M.Mio.(214,217,253,254,R).
Pullenia osloensis Feyling-Hanssen = *Pullenia quinqueloba* (Reuss) subsp. *minuta* Feyling-Hanssen, 1954, Nor.Geol.Tidsskr., 33(1—2): p.133, pl.2, f.3 [emend. *P. osloensis* Feyling-Hanssen, 1954, Nor.Geol.Tidsskr., 33(3—4): p.194]. Olig.—L.Mio. (214,F; 216,217,253,R); M.Mio.(214,216,217,F; 253,254,R); U.Mio.(214,216,217,R—F; 254,R); Plio.(254,R; 214,216,217,F—C); Quat.(214,R; 216,217,C).
The taxonomy as well as the morphology of this species were very well discussed by Feyling-Hanssen (*loc. cit.*). One detail, however, found in my specimens (and in those from the North Atlantic Ocean kindly mailed to me by A. Lamy) was not mentioned by Feyling-Hanssen. Many of my specimens have an asymmetrical last chamber with one side more developed than the other. This causes the margin of the last chamber to be located at an angle to the principal axis of the tests.
Pullenia quadriloba Reuss = *Pullenia compressiuscula* Reuss var. *quadriloba* Reuss, 1867, K.Akad. Wiss.Wien, Math-Naturw. Cl., S.-B., 55, (Abt.1): p.87, pl.3, f.8. Olig.(216,217,R); L.Mio.(214,216,R); M.Mio.—U.Mio.(214,253,254,R).
Pullenia salisburyi Stewart and Stewart, 1930, J.Paleontol., 4(1): p.72, pl.8, f.2. Olig.-Quat.(214,216,217,R); U.Mio.(253,R).
Pullenia subcarinata quinqueloba (Reuss) = *Nonionina quinqueloba* Reuss, 1851, Dtsch.Geol.Ges., Zeitschr., 3: p.71, pl.5, f.31. Olig.(214,216,217, 253,254,R); L.Mio.(214,253,254,R; 216,217,F); M.Mio.(214,216,217,253,254,R); U.Mio.(214,216,217,253,R); Plio.(214,253,254,R; 216,217,F); Quat.(216,254,R).
Boltovskoy (1959), after studying Recent material from the SW Atlantic, came to the conclusion that *P. quinqueloba* (Reuss) is a subspecies of *P. subcarinata* (d'Orbigny). Thus, they should be considered as *P. subcarinata subcarinata (nominat subspecies)* and *P. subcarinata quinqueloba*. The latter differs from the *nominat subspecies* mainly in having depressed sutures a lobulate peripheral outline and five (but not 5—6) chambers in the last whorl. An uninterrupted chain of transitional forms between both foraminifera was observed. Although in this report *P. quadriloba* and *P. salisburyi* are cited as independent species, it is evident that they are very closely related to *P. subcarinata* and probably are some of its variants also.

Pullenia subcarinata subcarinata (d'Orbigny) = *Nonionina subcarinata* d'Orbigny, 1839, Voy.Am.Mérid., 5(5): p.28, pl.5, f.23—24. L.Mio.(253,R); L.Mio.—U.Mio.(216,217,R); Plio.—Quat.(214,217,254,R).

Pyrgo depressa (d'Orbigny) = *Biloculina depressa* d'Orbigny, 1826, Ann.Sci.Nat., sér.1, 7: p.298, Mod.91. U.Mio.(217,253,R); Plio.—Quat.(214,R).

Pyrgo lucernula (Schwager) = *Biloculina lucernula* Schwager, 1866, "Novara" Exped., Geol., 2: p.202, pl.4, f.14. Plio.—Quat.(253,R).

Pyrgo murrhina (Schwager) = *Biloculina murrhina* Schwager, 1866, "Novara" Exped., Geol., 2: p.203, pl.4, f.15. M.Mio.(253,R); U.Mio.—Plio.(214,216,253,254,R); U.Mio.—Quat.(214,216,R; 217,253,F).

Pyrgo aff. *P. nasuta* Cushman, 1935, Smithson. Inst.Misc.Coll., 91(21): p.7, pl.3, f.1—4. Plio.(214,217,R).
The sign "aff." is used because the specimens recovered have a narrower peripheral keel and a shorter neck than typical *P. nasuta* specimens.

Pyrgo serrata (Bailey) = *Biloculina serrata* Bailey, 1861, Boston J.Nat.Hist., 7(3): p.350, pl.8, f.E. U.Mio.(217,R).

Pyrulina fusiformis (Roemer) = *Polymorphina fusiformis* Roemer, 1838, Neues Jahrb. Mineral.Geogr.Geol. Petrogr., p.386, pl.3, f.37. Olig.(253,R); L.Mio.—M.Mio. (254,R); U.Mio.(214,217,253,R); Plio.(214,R); Quat.(253,R).

Quinqueloculina cf. *Q. pygmaea* Reuss, 1850, K.Akad. Wiss.Wien, Math.-Nat. Cl., Denkschr., 1: p.384, pl.50, f.3. U.Mio—Quat.(217,R).
The specimens found more or less fit the description and figure of *Q. pygmaea*; they differ from it mainly in being somewhat narrower and in having a poorly developed neck.

Quinqueloculina venusta Karrer, 1868, K.Akad. Wiss.Wien, S.-B., 58(Abt.1): p.147, pl.2, f.6. M.Mio.(217,R); U.Mio.(217,253,R); Plio.(253,R).
The specimens of the Indian Ocean have on the average less sharp peripheral angles and in many cases the shape of the test is less elongated.

Quinqueloculina weaveri Rau, 1948, J.Paleontol., 22: p.159, pl.28, f.1—3. M.Mio.(216,253,R); U.Mio.(214,217,253,R); Plio.(214,253,254,R); Quat.(216,253, 254,R).
This species is subject to considerable morphological variations in its general outline, character of chambers and size. The aperture is also variable. Although Rau (loc. cit.) stated that this species does not show apparent dentition, several of my specimens have a poorly developed tooth.

Rectuvigerina royoi Bermúdez and Fuenmayor, 1963, in: Bermúdez and Seiglie, 1963, Inst.Ocean.Univ. Oriente, Bol., 2(2): p.144, pl.18, f.9. L.Mio.(214,253,R); M.Mio.—U.Mio.(214,253,254,R); Plio.(254,R).

Robulus cassis rotundatus (Silvestri) = *Cristellaria cassis* Lamarck in: d'Orbigny, 1846, Foram.Foss.Vienne, p.91, pl.4, f.4—7 = *Cristellaria cassis* (Fichtel and Moll) var. *rotundata* Silvestri 1898, Acad.Sci.Lett. Arti Acir., Cl.Sci., Atti Rend., Acireale, n.s., 8: pp.66,113,114. L.Mio.(253,R); M.Mio.(254,R).

Robulus rotulatus (Lamarck), s.l. = *Lenticulites (rotulata)* Lamarck, 1804, Ann.Mus., 5: p.188, no.3, pl.62, f.11 (1806). Olig.—M.Mio.(254,R—F); Olig.—U.Mio.(214,216, 217,253,R); Plio.(214,217,254,R); Quat.(214,217,253,R).
I prefer to interpret my specimens as sensu lato because there are many which show some morphological variation and perhaps would be identified by other foraminiferologists as other species close to *R. rotulatus*. They are connected by linkage forms and I think that it is best to ascribe all of them to the same species — *Robulus rotulatus*, s.l.

Saracenaria latifrons jamaicensis Cushman and Todd, 1945, Cushman Lab.Foraminiferal Res., Spec. Publ., 15: p.32, pl.5, f.7. Olig.(217,R); L.Mio.(254,R); M.Mio.(214,216,253,254,R); U.Mio.(214,253,R).

Sigmoilina schlumbergeri Silvestri, 1904, Accad.Pont. Romana N. Linc., Mem., 22: p.267, 269; Schlumberger, 1887, Boll. Soc. Fr., 12: pl.7, f.12—14. Plio.(214,R).

PLATE VII

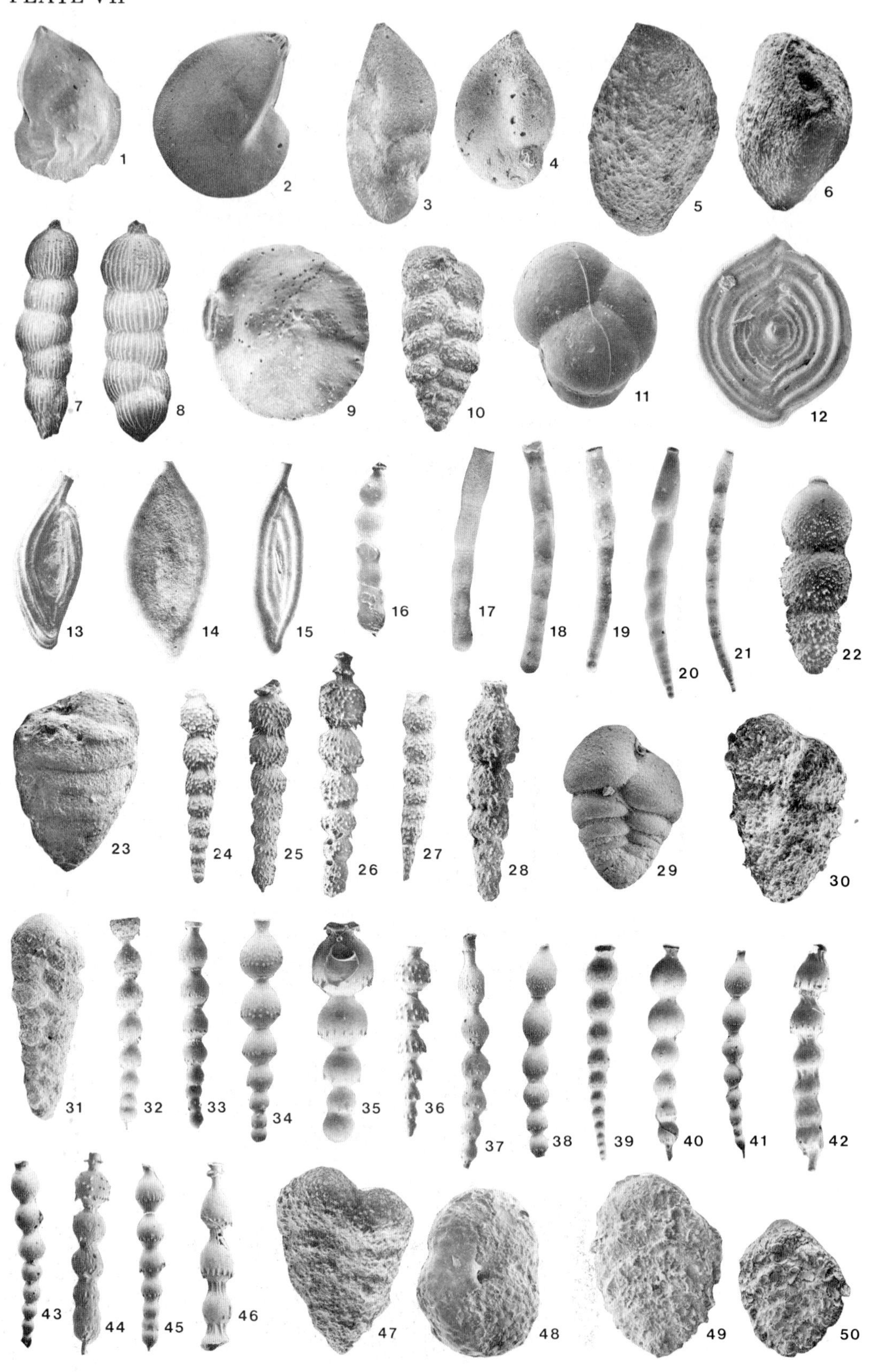
1
2
3
4
5
6
7
8
9
10
11
12
13
14
15
16
17
18
19
20
21
22
23
24
25
26
27
28
29
30
31
32
33
34
35
36
37
38
39
40
41
42
43
44
45
46
47
48
49
50

Siphogenerina vesca Finlay, 1939, Roy.Soc. N.Z., Trans., Proc., 69(part 1); p.109, pl.13, f.46,47. Olig.—M.Mio.(254,R).
Siphonina reticulata (Czjzek) = *Rotalina reticulata* Czjzek, 1848, Haidinger's Nat. Abhandl., 2: p.145, pl.13, f.7—8. Olig.(253,R); L.Mio.(214,R).
Siphotextularia rolshauseni Phleger and Parker, 1951, Geol.Soc.Am., Mem., 46(part 2): p.4, pl.1, f.23,24. Olig.(216,253,R); L.Mio.(214,216,253,R); M.Mio.(214,216,217, 253,254,R); U.Mio.—Plio.(214,216,217,R).
Sphaeroidina bulloides d'Orbigny, 1826, Ann.Sci.Nat., sér.1, 7: p.267, Mod.65. Olig. (214,216,R); L.Mio.(214,216,254,R); M.Mio.(214,217,253,254,R); U.Mio.(214,216, 217,254,R); Plio.(214,216,217,253,254,R); Quat.(214,216,254,R).
Spiroloculina antillarum d'Orbigny, 1839, in: de la Sagra, Hist.Phys. Nat. Cuba. Foram., p.166, pl.9, f.3,4. Quat.(254,R).
Spiroloculina asperula Karrer, 1868, K.Akad. Wiss.Wien, S-B., 58(Abt.1): p.136, pl.1, f.10. M.Mio.—Plio.(254,R).
Spiroloculina esnaensis LeRoy, 1953, Geol.Soc.Am., Mem., 54: p.49, pl.3, f.15,16. Olig.(254,R).
Spiroloculina pusilla Earland, 1934, "Discovery" Rep., 10: p.47, pl.1, f.3,4. M.Mio. (217,R); Quat.(217,R).
Stilostomella abyssorum (Brady) = *Nodosaria abyssorum* Brady, 1884, "Challenger" Exped., Rep., Zool., 9: p.504, pl.63, f.8,9. Olig.(214,216,F; 217,253,R); L.Mio.(214,216,253,254,R); M.Mio.(214,217,R); U.Mio.(214,216,217,253,R); Plio.(214,R).
Stilostomella cf. *S. annulifera* (Cushman and Bermúdez) = *Ellipsonodosaria annulifera* Cushman and Bermúdez, 1936, Cushman Lab.Foraminiferal Res., Contrib., 12(part 2):

PLATE VII

1. *Robulus cassis rotundatus* (Silvestri); × 20; site 253.
2. *Robulus rotulatus* (Lamarck), s.l.; × 25; site 254.
3, 4. *Saracenaria latifrons jamaicensis* Cushman and Todd; × 45; site 254.
5, 6. *Sigmoilina schlumbergeri Silvestri*; × 30; (fig.5), × 50 (fig.6, apertural view); site 254.
7, 8. *Siphogenerina vesca* Finlay; × 45; site 254.
9. *Siphonina reticulata* (Czjzek); × 50; site 253.
10. *Siphotextularia rolshauseni* Phleger and Parker; × 70 (fig.10); site 253.
11. *Spaeroidina bulloides* d'Orbigny; × 30; site 253.
12. *Spiroloculina esnaensis* LeRoy; × 20; site 254.
13. *Spiroloculina antillarum* d'Orbigny; × 40; site 254.
14. *Spiroloculina asperula* Karrer; × 45; site 254.
15. *Spiroloculina pusilla* Earland; × 65; site 217.
16. *Stilostomella abyssorum* (Brady); × 5; site 216.
17—21. *Stilostomella* cf. *S. annulifera* (Cushman and Bermúdez); × 45 (figs.17—19), × 25 (figs.20,21); site 253 (figs.17,19,20), site 254 (figs.18,21).
22. *Stilostomella tuckerae* (Hadley); × 25; site 216.
23. *Textulariella goesii* (Cushman); × 20; site 254.
24—28. *Stilostomella subspinosa* (Cushman); × 15; site 253.
29. *Textularia flintii* Cushman; × 25; site 253.
30. *Textularia kapitea* Finlay; × 35; site 254.
31. *Textularia porrecta* Brady; × 50; site 254.
32—46. *Stilostomella* ex gr. *S. lepidula* (Schwager); × 25 (figs.32,33,36,40,43), × 20 (figs.38,39,41,42,44—46), × 35 (figs.35,37), × 45 (fig.34); site 253 (figs.32,33,36,38—46), site 254 (fig.34), site 217 (figs. 35, 37).
47, 48. *Textularia halkyardi* Lalicker; × 50 (fig.47), × 40 (fig.48, apertural view); site 254.
49, 50. *Textularia milletti* Cushman; × 30; site 253.

p.28, pl.5, f.8,9. Olig.(217,R; 254,C); L.Mio.(214,253,R; 254,C); M.Mio.—Plio.(214, 217,253,R; 216,254,F—C); Quat.(214,253,R).

The closest species to my specimens is this one of Cushman and Bermúdez. The following differences force me to identify them with "cf." Typical *S. annulifera* specimens have somewhat shorter chambers and a much better pronounced raised annular appearance of the sutures. Based on their external morphology, my specimens have much in common with *Dentalina filiformis* and *D. costai*. However, they do not have a radiate aperture and are typical *Stilostomella*.

Stilostomella ex gr. *S. lepidula* (Schwager) = *Nodosaria lepidula* Schwager, 1866, "Novara" Exped., Geol., 2: p.210, pl.5, f.27,28. Olig.—U.Mio.(214,216,217,253,254, R—C); Plio.(214,253,F; 217,254,R); Quat.(214,253,254,R).

Probably no foraminifer in the whole Indian Ocean material took so much time and effort to identify than this one and I am still not completely satisfied with my decision to call it *Stilostomella* ex gr. *S. lepidula*. However, for the time being, this decision seems the best one to me.

This species is subject to great morphological variation in several of its traits. The most conspicuous feature, the presence of spines situated around the lower part of each chamber, is probably the most changeable; these spines may form a line, two lines or may be just scattered. In many specimens they are practically invisible in a common microscope and only SEM micrographs reveal their presence. In some specimens they are absent in the first portion of the test. Other changeable traits are the character of the separation between chambers, the shape of the chambers, the presence or absence of the initial spines on the proloculus, etc. Based on their aperture, the specimens discussed should be ascribed to *Stilostomella*, because many specimens possess a tooth. However, some tests were found without a tooth. It is interesting to note that no other morphological differences could be observed in many cases between the tests with a tooth and those without it; therefore it seems that the difference in the aperture of the genera *Stilostomella* and *Orthomorphina* has relatively little value.

In view of the great quantity of species (described often under different generic names) which very well fit my specimens, it would not be difficult to separate them into several species. I think, however, that this would be erroneous, because in the fauna under study all the specimens are connected with transitional forms and therefore to recognise distinct species would be a mistake.

For these reasons I consider them as one species interpreting them in broad sense as *Stilostomella* ex gr. *S. lepidula*.

Stilostomella subspinosa (Cushman) = *Ellipsonodosaria subspinosa* Cushman, 1943, Cushman Lab. Foraminiferal Res., Contrib., 19(part 4): p.92, pl.16, f.6,7. Olig.—U.Mio.(214,216,217,R; 253,F); Plio.—Quat.(253,R).

Stilostomella tuckerae (Hadley) = *Ellipsonodosaria tuckerae* Hadley, 1934, Bull. Am. Paleontol., 20(70A): p.21, pl.3, f.1,2. Olig.(214,216,R); L.Mio.—M.Mio.(214,216,217, 253,R); U.Mio.(214,253,R).

Textularia flintii Cushman, 1911, U.S.Natl.Mus., Bull., 71(part 2): p.21, tf.36. L.Mio. (216,R); M.Mio.(214,216,254,R); U.Mio.(216,254,R); Plio.(214,254,R); Quat.(214,R).

Textularia goesii Cushman, 1911, U.S.Natl.Mus., Bull., 71(part 2): p.15, tf.24. L.Mio.—M.Mio.(254,R).

Textularia halkyardi Lalicker, 1935, Cushman Lab. Foraminiferal Res., Contrib., 11: p.45, pl.7, f.5. Olig.(214,R); L.Mio.(253,254,R); M.Mio.(214,R); U.Mio.(253,R); Plio.—Quat.(254,R).

Textularia kapitea Finlay, 1947, N.Z. J. Sci. Technol., Ser.B, 28(5): p.266, pl.2, f.21,22. Olig.—L.Mio.(254,R).

Textularia milletti Cushman, 1911, U.S.Natl.Mus., Bull. 71(part 2): p.13, tf.18,19. L.Mio.(253,R); M.Mio.(254,R); U.Mio.(214,253,R); Plio.(214,253,254,R); Quat.(254,R).

Textularia porrecta Brady = *Textularia agglutinans* d'Orbigny, var. *porrecta* Brady, 1884. "Challenger" Exped. Rep., Zool., 9: p.364, pl.43, f.4. Plio.—Quat.(253,R).

Trifarina bradyi Cushman, 1923, U.S.Natl.Mus., Bull., 104(part 4): p.99, pl.22, f.3—9. L.Mio.—U.Mio.(254,R—F).

Uvigerina aculeata d'Orbigny, 1846, Foram.Foss. Vienne, p.191, pl.11, f.27,28. U.Mio.(214,R).

Uvigerina graciliformis Papp and Turnovsky, 1953, Geol. Bundesanst., Jahrb. 96(1): p.122, pl.5, f.5—7. Olig.(216,217,253,R); L.Mio.(214,216, 217, 253, 254,R); M.Mio.(214,217,254,R; 216,F); U.Mio.(214,217,253,R; 216,F); Plio.(253,R).
This species was described as *Uvigerina*. It could be interpreted as *Hopkinsina* as well, because its last five-six chambers, according to Papp and Turnovsky (ut supra), have both bi- and uniserial chambers at the apertural end. It is a very variable species, especially as far as its sculpture is concerned. My specimens resemble *U. graciliformis* sufficiently well, but there are some differences, for instance, they do not form point-like prolongations of the ribs as some typical representatives of *U. graciliformis* do. Nevertheless, these differences are not of great importance and fall into the category of infraspecific variations.

Uvigerina hispida Schwager, 1866, "Novara" Exped., Geol., 2: p.249, pl.7, f.95. M.Mio.—U.Mio.(254,R—F).

Uvigerina miozea Finlay, 1939, R.Soc. N.Z., Trans., Proc., 69(part 1): p.102, pl.12, f.12—14. Olig.—U.Mio.(214,253,R—F).
There are some differences between my specimens and those described by Finlay. However, they are not of taxonomical importance. According to Finlay, all the chambers are ribbed, but the last few chambers of my specimens are not ribbed. Moreover, the specimens from the present material have a relatively longer apertural neck.

Uvigerina peregrina Cushman, 1923, U.S.Natl.Mus., Bull., 104(part 4): p.166, pl.42, f.7—10. Olig.—L.Mio.(216,R); M.Mio.(217,253,R); U.Mio.(216,F; 253,R); Plio.(216, 253,254,R); Quat.(214,216,217,253,254,R—F).
The sculpture of this species is very variable and it is quite probable that many of my specimens would be ascribed by other researchers to *U. peregrina dirupta* Todd.

Uvigerina porrecta Brady, 1879, Q.J.Microsc.Sci., new ser., 19: p.60, pl.8, f.15,16. L.Mio.—M.Mio.(254,R).

Uvigerina proboscidea Schwager, s.l. 1866, "Novara" Exped., Geol., 2: p.250, pl.7, f.96. Olig.—M.Mio.(214,216,253,254,R—F); U.Mio.—Plio.(214,216,C; 217,253,254,R—F); Quat.(214,216,C; 253,254,R).
Finely hispid varieties of *Uvigerina* are common dwellers of great depths in Recent and Neogene oceans. They have been described many times under several names. The oldest and the most used are *U. auberiana*, *U. asperula* and *U. proboscidea*. I prefer to call my specimens *U. proboscidea*, s.l. because *U. auberiana* has chambers arranged in two opposed alternating series (and thus they are not real *Uvigerina* specimens). The spines of *U. asperula* specimens should be arranged in more or less distinct rows. In my specimens the spines are scattered and thus the specimens fit well only the description of *U. proboscidea*.
This is very variable species and in the population found it is quite easy to differentiate several variants (I prefer to call them forms) which have been previously described as different taxa. Among these forms the most common is *U. proboscidea*, f. pseudo-ampullacea (described by Asano as *U. pseudoampullacea*, n. sp.). Isolated specimens of both f. ampullacea (described by Brady as *U. asperula*, var. ampullacea, n. var.) and f. vadescens (described by Cushman as *U. proboscidea*, var. vadescens, n. var.) were found too. Even some specimens which show a clear tendency toward the species described by Brady under the name *U. interrupta* were encountered in the material studied. All these forms (connected with linkage forms) were found in the same samples and cannot be considered as independent taxa.

Uvigerina schencki Asano, 1950, in: L. W. Stach (Editor), Illustr. Catalog. Jpn. Tert. Small. Foram., Tokyo, Hosokawa, p.17, tf.74—75. U.Mio.(253,254,R); M.Mio.—L.Mio.(254, I—R).

PLATE VIII

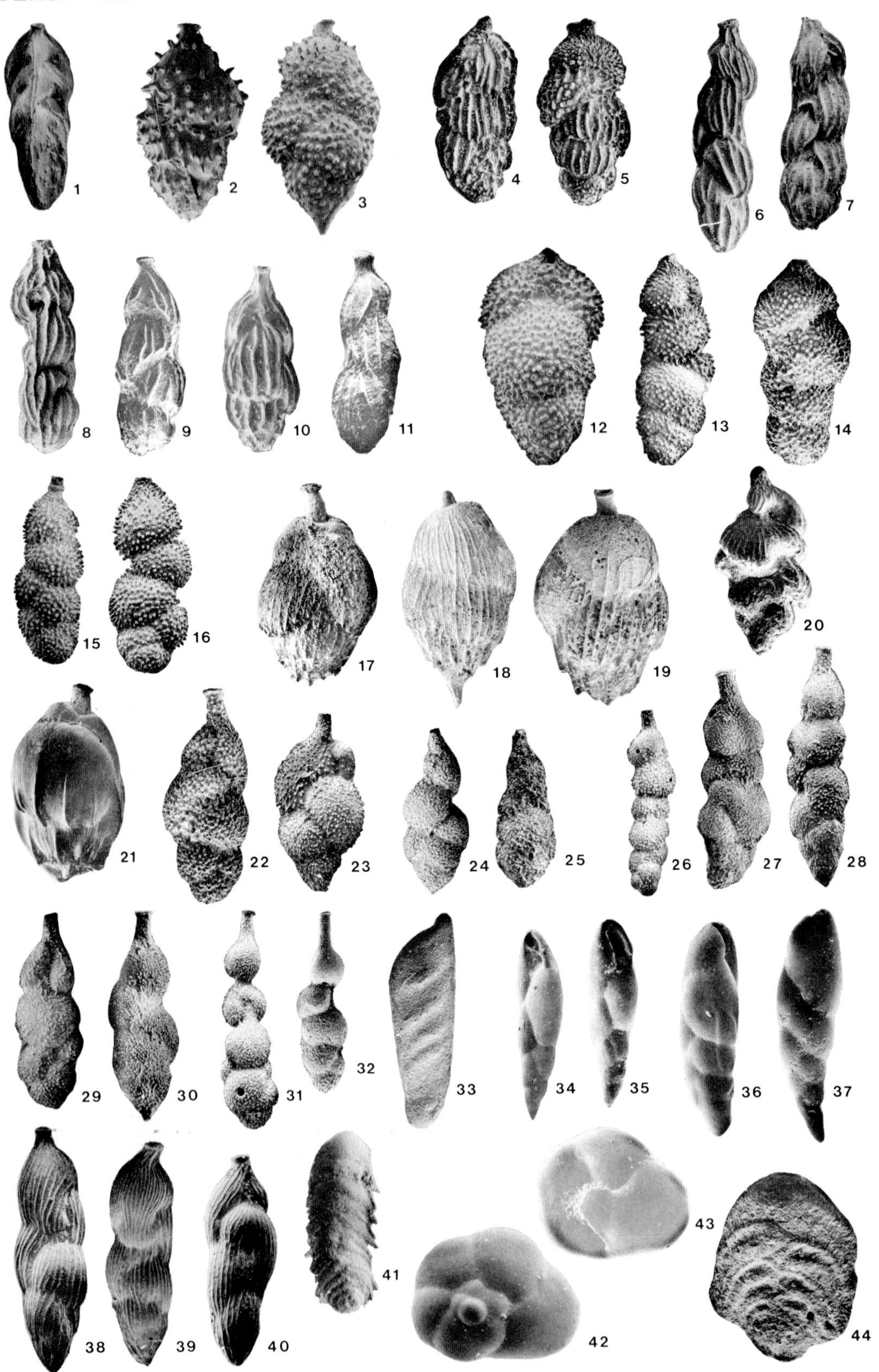
1
2
3
4
5
6
7
8
9
10
11
12
13
14
15
16
17
18
19
20
21
22
23
24
25
26
27
28
29
30
31
32
33
34
35
36
37
38
39
40
41
42
43
44

Uvigerina schwageri Brady, 1884, "Challenger" Exped., Repts., Zool., 9: p.575, pl.74, f.8—10. M.Mio.(254,R); Plio.(254,R).
Vaginulina legumen (Linnaeus) = *Nautilus legumen* Linnaeus, 1758, Syst. Nat., ed.10: p.711, no.248; 1767, Syst.Nat., ed.12: p.1164, no.288. Olig.—L.Mio.(254,R).
Valvulineria laevigata Phleger and Parker, 1951, Geol.Soc.Am., Mem., 46(part 2): p.25, pl.13, f.11,12. U.Mio.(214,216,R).
Virgulina complanata Egger, 1893, "Gazelle" Exped., K. Bayr.Akad. Wiss., Abh., Cl.II, 18: p.292, pl.8, f.91,92. M.Mio.(216,217,R); Plio.(254,R); Quat.(216,217,R).
Virgulina primitiva Cushman, 1936, Cushman Lab. Foraminiferal Res., Spec.Publ., 6: p.46, pl.7, f.1. Olig.—U.Mio.(214, 216,253,254,I-*f*); Plio.(254,I).
Virgulina texturata Brady, 1884, "Challenger" Exped., Rep., Zool., 9: p.415, pl.52, f.6. M.Mio.—U.Mio.(254,R).
Vulvulina pennatula (Batsch) = *Nautilus (Orthoceras) pennatula* Batsch, 1791, Conchyl. Seesandes, pl.4, f.13. Olig.—L.Mio.(214,R; 216,253,254,R—F); M.Mio.(214,253, 254,R); U.Mio.—Plio.(214,253,254,R).
?*Vulvulina spinosa* Cushman, 1927, Cushman Lab. Foraminiferal Res., Contrib., 3(part 2): p.111, pl.23, f.1. Olig.(214,216,253,254,R); L.Mio.—U.Mio.(214,253,R).
This identification is tentative because the single difference of importance between these specimens and *V. pennatula* are presence of the spinose projections at the basal peripheral angle. However, it is questionable if this difference can be used as species character.

MAIN CONCLUSIONS

(1) The whole Upper Oligocene—Quaternary sequence studied on the Ninetyeast Ridge is dominated by a few species which occur throughout the sequence. This is evidence that no appreciable change in depth has taken place since Late Oligocene time in the area of the Ninetyeast Ridge.

PLATE VIII

1. *Trifarina bradyi* Cushman; × 45; site 254.
2, 3. *Uvigerina aculeata* d'Orbigny; × 35 (fig.3), × 20 (fig.4); site 214.
4, 5. *Uvigerina peregrina* Cushman; × 30; site 253.
6—11. *Uvigerina graciliformis* Papp and Turnovsky; × 105 (figs.6,10), × 70 (figs.7,8), × 75 (figs.9,11); site 216 (figs.6,11); site 254 (figs.9,10); site 253 (figs.7,8).
12—16. *Uvigerina hispida* Schwager; × 45 (fig.12), × 25 (figs.13—16); site 254 (fig.12); site 253 (figs.13—16).
17—19. *Uvigerina miozea* Finlay; × 35 (figs.17,18), × 40 (fig.19); site 253.
20. *Uvigerina porrecta* Brady; × 50; site 254.
21. *Uvigerina schwageri* Brady; × 35; site 254.
22, 23. *Uvigerina proboscidea* Schwager, forma typica; × 50; site 253.
24, 25. *Uvigerina proboscidea* Schwager, forma ampullacea Brady; × 40; site 253.
26—28. *Uvigerina proboscidea* Schwager, forma pseudoampullacea Asano; × 50 (fig.26), × 40; (fig.27), × 35 (fig.28); site 253.
29, 30. *Uvigerina proboscidea* Schwager, forma vadescens Cushman; × 50; site 254.
31, 32, *Uvigerina proboscidea* Schwager, transitional form to forma interrupta Brady; × 30; (fig.31), × 35 (fig.32); site 254.
33. *Vaginulina legumen* (Linnaeus); × 25; site 254.
34, 35. *Virgulina complanata* Egger; × 60; site 254 (fig.34); site 217 (fig.35).
36, 37. *Virgulina texturata* Brady; × 35; site 254.
38—40. *Uvigerina schencki* Asano; × 45; site 254.
41. ?*Vulvulina spinosa* Cushman; × 8; site 253.
42, 43. *Valvulineria laevigata* Phleger and Parker; × 70; site 214.
44. *Vulvulina pennatula* (Batsch); × 25; site 253.

(2) A few benthonic foraminiferal species, which have limited time ranges and thus can be used as guide fossils, were encountered in the material studied. Several of these species occur only rarely. The majority of benthonic foraminifera found in the present study apparently have long time ranges.

(3) It was possible to locate by means of benthonic foraminifera only two stratigraphic boundaries, namely the Middle Miocene/Upper Miocene and the Upper Miocene/Pliocene. At other boundaries, even the Oligocene/Lower Miocene one, benthonic foraminiferal fauna do not show changes of importance.

(4) Interesting evolutionary trends were observed in two cases: *Cibicides wuellerstorfi* originated from *Planulina marialana gigas* and various *Cassidulina* (mainly *C. subglobosa subglobosa*) originated from *C. cuneata.*

(5) The range chart (Table III) of selected species (the most numerous and consistent ones, as well as those which have a limited time range) gives an idea about the general aspect of foraminiferal benthonic assemblages of different units and their guide fossils in the Late Cenozoic sequences of the Ninetyeast Ridge.

(6) A comparison of the results obtained by this study and that by a study of the Late Cenozoic Pacific deep-water benthonic foraminifera (Douglas, 1973) showed that only a very limited number of species can be used for worldwide stratigraphic purposes (Table IV).

(7) Deep-water benthonic foraminifera have limited value as time indicators and can be used as such mainly for determination of major stratigraphic sections. However, they are very useful in deciphering several ecological parameters as well as for the recognition of re-deposited material. Therefore it is highly desirable that a study of benthonic foraminifera is permanently included in the paleontological studies of deep sea drilling cores.

ACKNOWLEDGEMENT

I express my sincere thanks to the following institutions and persons:

(1) The authorities of the Deep Sea Drilling Project for their kind invitation to participate onboard the "Glomar Challenger" for Leg 26 as well as for the samples of Leg 22 sent to me for this study.

(2) Mr. J. Remiro, Mrs. I. Riobó de Magaldi and Miss A. M. Leverone for their technical help in preparation of both material and manuscript.

(3) The staff of the Scanning Electron Microscope Service of the CONICET (Argentina) for taking all the photomicrographs.

(4) Dr. R. Cifelli for critically reviewing the manuscript and making several suggestions.

REFERENCES

Boltovskoy, E., 1965. Twilight of foraminiferology. J.Paleontol., 39: 383–390.
Boltovskoy, E., 1974. Neogene planktonic foraminifera of the Indian Ocean,